D0812601

Spin Labeling Methods
in Molecular Biology

Spin Labeling Methods in Molecular Biology

(Metod Spinovykh Metok v Molekulyarnoi Biologii)

G. I. LIKHTENSHTEIN

Institute of Chemical Physics,
USSR Academy of Sciences
(Izdatel'stvo Nauka, Moscow, 1974)

Translated by Philip S. Shelnitz

A Wiley-Interscience Publication

JOHN WILEY & SONS

New York   London   Sydney   Toronto

2   UDC 539.143.42:577.23

Spin Labeling Methods in Molecular Biology, G. I. Likhtenshtein, Nauka, Moscow, 1974, 256 pages.

This book presents the theoretical and experimental basis for the application of spin labels and probes, that is, the methods for preparing specifically modified spin-labeled proteins, enzymes, nucleic acids, and other biopolymers, the theoretical principles underlying the EPR spectra of nitroxide radicals and paramagnetic ions, and the methods for treating experimental EPR spectra. The possible uses of the method are the subjects of special sections. These include determination of distances between spin labels (including those in the active centers of enzymes), investigation of conformational changes in proteins, nucleic acids, and membranes under various effects, detection of fine generalized concerted transitions and allosteric effects in enzymes, and study of the molecular and conformational mobility of various biological structures. The monograph is intended for scientists working in various areas of molecular biology, biophysicists, and biochemists, as well as the instructors, graduate students, and undergraduates of university biology departments.

Tables, 17.   Figures, 85.   Bibliography, 583 entries.

A. L. Buchachenko, D. Chem. Sc., editor.

[© Izdatel'stvo Nauka, 1974.]

All Rights Reserved.
Authorized translation from Russian language edition published by Izdatel'stvo Nauka.

Copyright © 1976 by John Wiley & Sons, Inc.

All rights reserved. Published simultaneously in Canada.

No part of this book may be reproduced by any means, nor transmitted, nor translated into a machine language without the written permission of the publisher.

Library of Congress Cataloging in Publication Data:

Likhtenshtein, Gerts Il'ich.
   Spin labeling methods in molecular biology =
Metod spinovykh metok v molekulyarnoi biologii.

   Includes bibliographical references and index.
   1. Spin labels.   I. Title.

QH324.9.S62L5413   574.8'8'028   76-16500
ISBN 0-471-53415-3

Printed in the United States of America

10 9 8 7 6 5 4 3 2 1

Preface

This monograph treats the principles of the spin-labeling technique and the results of applying this method in contemporary molecular biology. The study of labeled radicals of biological compounds by electron paramagnetic resonance (EPR) has made it possible to follow the motion of a paramagnetic center, to determine the distance between labels added to specific groups, to probe the polarity and mobility of molecular environments, and to detect changes in structures as they function or when certain defects act on them.

One of the essential first steps in creating the spin-labeling method was the work on the synthesis of a new class of piperidine and pyrrolidine stable nitroxide radicals. This research was carried out in the 1959–1963 period at the Institute of Chemical Physics of the USSR Academy of Sciences by Professor É. G. Rozantsev and his co-workers, under the initiative of Professor M. B. Neiman. Their discovery of a reaction that does not affect the free valence is especially important. The study of the physical properties of nitroxide radicals carried out at the Institute of Chemical Physics by Professor A. L. Buchachenko and his co-workers and outside the Soviet Union by McConnell's and Kivelson's groups laid the physical foundation for the use of spin labels. Ten years ago, while undertaking basic research in the area of enzymology and becoming acquainted with this amazing world of specificity and high rates, we arrived at the conclusion that nitroxide radicals can be used to determine the relative arrangement of functional groups in enzymes and to study local conformational changes and the dynamic structure of proteins. However, several years passed before we succeeded in overcoming the technical and psychological difficulties involved in rendering the new method practical.

The birthday of spin labeling can be placed at the end of 1965, with the report by Professor McConnell of Stanford University and his co-workers on the synthesis and investigation of the first spin-labeled preparations of bovine serum albumin. McConnell also made an important contribution to the subsequent development of the method.

Many new possible applications of spin labels are based on investigations carried out in the Soviet Union. This is primarily true of work on the chemistry of stable nitroxide radicals carried out under the direction of Professor É. G. Rozantsev (V. A. Golubev, Yu. V. Kokhanov, L. A. Krinitskaya, V. I. Suskina, A. B. Shapiro, et al.), on the investigation of the rotational and translational diffusion of radicals in different model systems (A. L. Buchachenko, A. M. Vasserman, V. B. Stryukov, et al.), and of the theoretical investigations in the field of slow and anisotropic motions (I. V. Aleksandrov, A. L. Buchachenko, N. N. Korst, A. N. Kuznetsov, A. V. Lararev, V. B. Stryukov, L. I. Antsifirova, et al.). Soviet scientists were the first to propose and develop such general approaches as the double spin-label technique (G. I. Likhtenshtein, P. Kh. Bobadzhanov, E. N. Frolov, A. V. Kulikov, et al.) and the "spin probe–spin label" method (G. I. Likhtenshtein and Yu. B. Grebenshchikov), to observe generalized concerted transitions in enzymes without a quaternary structure (G. I. Likhtenshtein, Yu. D. Akhmedov, B. P. Atanasov, M. V. Vol'kenshtein, V. P. Timofeev, O. L. Polyanovskii, et al.), and to study the local mobility of the different layers of water-protein enzyme matrices (G. I. Likhtenshtein, T. V. Avilova, et al.) and the structure of the active centers of a number of complex enzyme systems (G. I. Likhtenshtein, E. N. Frolov, A. V. Kulikov, L. M. Raikhman, et al.). The first work on the probing of nucleic acids and biological membranes with nitroxide radicals was also carried out in the Soviet Union (B. I. Sukhorukov, V. K. Kol'tover, et al.). Soviet scientists have proposed original techniques for studying the structure and function of biological membranes, nucleic acids, and other substances (A. É. Kalmanson, A. N. Kuznetsov, V. K. Kol'tover, L. M. Raikhman, and G. L. Grigoryan).

Our attention has been concentrated on general aspects of the practical application of spin labels, including the methods developed at the Laboratory for the Kinetics of Enzyme Action of the Institute of Chemical Physics of the USSR Academy of Sciences (Chapters 1–5), and on various possible uses of the method for solving certain problems in enzymology (Chapters 6, 7, and 9). Some very promising methods of investigation based on the combined use of EPR and nuclear magnetic resonance (NMR) are presented in Chapter 5. In Chapter 9 there is a discussion of several general aspects of enzyme action with respect to the new data obtained with spin labels. In Chapter 8 there is a brief account of methods employing luminescent labels, luminescence-quenching labels, Mössbauer (γ-resonance) labels, and electron-scattering labels. These methods provide information on the structure and local mobility of biological structures that can be compared with the results obtained with spin labels. The possibilities and results of using spin labels and probes in the study of biological membranes and other biological structures are presented in Chapter 10, which was written by V. K. Kol'tover (the section entitled "Nucleic acids" was written by the author).

As a result of the small size of the monograph, the fact that nonSoviet reviews on the application of spin labels in biology have been published, and the existence of detailed monographs and reviews on the theory of the

EPR spectra of paramagnetic substances, including nitroxide radicals, we
have limited ourselves to a presentation of only the elements of the theory
of EPR spectra needed to intelligently use the method and extract information from experimental EPR spectra. We have intentionally avoided extensive descriptions of the methods and the specific results of their application
to biological and model systems that have been thoroughly treated in
widely circulated reviews and monographs.

We take this opportunity to recall the memory of our teacher, Professor
M. B. Neiman, who was the first to foresee the wide and diverse use of
nitroxide radicals.

We express our sincerest graditude to our co-workers from the Laboratory
for the Kinetics of Enzyme Action: T. V. Avilova, who participated in
writing part of Chapter 9 and L. V. Romanshina, A. V. Kulikov, E. N.
Frolov, Yu. B. Grebenshchikov, V. K. Kol'tover, L. A. Syrtsova, and
V. A. Gromoglasova for their great help in writing and formulating the
monograph.

Finally, we are extremely grateful to A. L. Buchachenko, G. N. Bogdanov, A. M. Vasserman, K. I. Zamaraev, A. N. Kuznetsov, V. B. Stryukov,
A. P. Pivovarov, O. L. Polyanovskii, and A. E. Shilov, who reviewed individual parts of the manuscript and offered important critical comments.

We believe that the effort involved in writing this monograph will have
been worthwhile if it arouses interest in spin labeling and makes it easier
for chemists, biochemists, and biophysicists to understand the principles
of this method to such an extent that they can use it to solve their own
problems.

G. I. LIKHTENSHTEIN

Moscow, Russia
January 1976

Introduction

The problems and questions facing contemporary molecular biology are extremely diverse. At the present time there are about 800 known enzymes or biological catalysts that effectively accelerate a large number of chemical reactions of various types. Together with the structural proteins, membranes, nucleic acids, hydrocarbons, and other compounds, enzymes form supramolecular formations in various and sundry combinations. Naturally, the structure and mechanism of biological systems can be studied only by calling on the entire arsenal of biochemical, physical, and chemical methods.

Important advances have been achieved in the past decade in research on the structure and function of various biological systems, including proteins. The development of methods for obtaining protein and enzyme preparations of high purity has considerably enhanced the effectiveness of direct physical and chemical methods for studying individual protein substances.

The most accurate and unambiguous method for investigating biological systems is X-ray diffraction analysis. Thanks to the brilliant development of this method, the three-dimensional structures of hemoglobin, lysozyme, chymotrypsin, ribonuclease, carboxypeptidase, and many other proteins have been deciphered.

Optical rotatory dispersion, adsorption, spectral and luminescence methods, as well as hydrodynamic techniques, have been employed to resolve various structural and dynamic questions. Despite the great effectiveness of these methods, they all have certain limitations. X-Ray diffraction analysis is applicable only to single-crystal preparations of comparatively low-molecular proteins. Many methods for studying proteins and other biological agents in solution yield only averaged values for the properties of these agents.

The search thus continues for new experimental techniques that would make it possible to draw a more detailed and precise picture of the arrangement of the different sections and groups in biological structures of various degrees of complexity and to monitor changes in this arrangement as they function, as well as to study the local mobility of different portions of biological matrices.

New methods for investigating biological structures based on the introduction of various labels into these structures [1–51] can be called on to solve the problems just cited. The basic idea underlying this method is the chemical modification of selected functional groups by special compounds (labels) whose properties enable monitoring of the state of biological matrices by physicochemical methods.

Four types of compounds are used as labels:

1. Paramagnetic centers (stable nitroxide radicals and ions with unpaired electrons) [1–10].
2. Luminescent, luminescence-quenching, and other chromophores [11–14].
3. Mössbauer atoms (^{57}Fe), which produce nuclear γ-resonance spectra [15,16].
4. Electron-scattering atomic groupings (polymercurials) that effectively scatter electrons and produce clear images on electron photomicrographs [17,18].

The problems that are solved by label methods can be divided into three types:

1. Determination of distances between functional protein groups and other groups, including those in the active centers of enzymes [1–7, 12–14, 20–24] and evaluation of the depth to which paramagnetic centers are embedded in biological matrices [24]).
2. Study of the local topography and mobility of different layers in water-protein, membranous, and other matrices in the vicinity of bound labels [3,4,7,25], as well as the dynamic state of active centers in enzymes [26–27].
3. Detection of conformational changes in proteins, enzymes, and other structures, including effects at a considerable distance from the active centers (allosteric effects and generalized concerted transitions) [1,3,4, 28–33].

The methods based on the use of labels are considerably less accurate than X-ray diffraction analysis; nevertheless, they have many advantages, such as:

1. The possibility of working with dilute aqueous solutions of proteins enzymes, and other systems.
2. The possibility of obtaining information on the dynamic characteristics of different portions of a biological matrix.
3. The possibility of studying the structure of active centers in comparatively high-molecular enzymes, including those appearing in complex biological structures.
4. Speed and greater accessibility to measurement.

New methods, of course, do not replace those already known that have become classical. However, their application in certain cases makes it possible to obtain unique information about the structure of the objects under investigation, and in other cases they can serve as independent methods for checking the results obtained by traditional methods.

The possibilities of using labels are especially evident in the study of enzyme action. In fact, the contemporary theories attribute the unique catalytic properties of enzymes to their multifunctional nature and dynamic structure. Determination of distances between functional groups in enzymes and detection of conformational changes and local mobility in matrices are necessary stages in the investigation of every enzyme.

The physicochemical properties of labels, that is, paramagnetic particles, luminescent chromophores, Mössbauer atoms, and electron-scattering groups, added to specific functional groups are altered under specific conditions by the respective compounds to which the labels are added. This makes it possible to determine distances between labels and active centers in enzymes when the latter contain paramagnetic or chromophoric groups. Knowledge of these distances and the chemical structure of the labels permits us to evaluate the distances between the groups to which the labels were added. The accuracy of such an evaluation is generally low (4–5 Å). However, in many cases it is adequate for resolving the question of whether the protein groups under investigation can undergo a direct (isosteric) interaction or whether the observed relationship between the groups is a long-range (allosteric) effect. Such questions are often encountered in deciphering the structure of enzyme centers and in investigating allosteric effects and generalized concerted transitions.

The physical parameters of labels (spin, luminescent, and Mössbauer) reflect the local viscosity and rigidity of biological matrices. This makes it possible to study the dynamic structure of the active centers of enzymes and to ascertain the degree of change in these structures under specific effects. Implanting labels on a protein macromolecule outside of the active center and, in particular, on other macromolecules appearing in the biological assemblage under investigation creates various interesting possibilities for studying long-range effects, which are apparently of primary importance in the expression and regulation of the catalytic properties of enzymes.

Several spin-labeling techniques, which are described in detail in Chapters 1–8 of the monograph, were proposed and developed mainly for application to the solution of problems related to enzyme action. However, they may possibly be of use in the study of the structure and dynamic properties of membranes, nucleic acids, carbohydrates, and other biological objects.

Contents

Chapter One
Effect of Rotational Diffusion on the EPR Spectra of Nitroxide Radicals

In the overwhelming majority of stable nitroxide radicals used as spin labels and probes, the nitroxide group is stabilized by substituents on the α-carbon atoms:

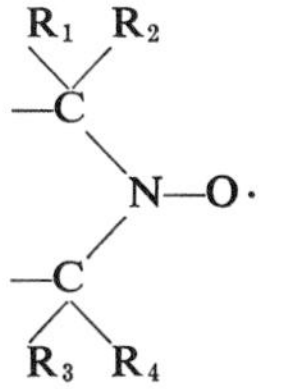

$$
\begin{array}{c}
R_1 \quad R_2 \\
\diagdown \diagup \\
-C \\
\diagdown \\
N{-}O\cdot \\
\diagup \\
-C \\
\diagup \diagdown \\
R_3 \quad R_4
\end{array}
$$

The nitroxide groups of stable radicals produce simple, well-resolved, and comparatively easily treated EPR spectra that are sensitive to molecular motion and to the structure of the local molecular environment.

Experimentally, the EPR spectra of nitroxide radicals are relatively weakly dependent on the nature of the substituents but strongly dependent on the state of the environment in which the radical is placed and on the presence of paramagnetic groups in this environment. In fact, the three spectral lines of almost identical width and intensity for dilute solutions of the radicals in low-viscosity solvents (water, methanol, and acetone) at room temperature (Fig. 1) begin to undergo abrupt transformations on transition to viscous media (glycerol) and, to an even greater extent, to solid matrices (frozen and supercooled solutions). The spectral lines then broaden, become unsymmetrical, and move increasingly further from each other.

Increasing the concentration of liquid and solid nitroxide solutions and adding other paramagnetic particles also cause dramatic changes in the EPR spectra of nitroxides; these changes, as seen in Fig. 1, are dependent on the experimental conditions. The contemporary theory predicts these changes and provides a qualitative and often quantitative interpretation for them.

ELEMENTS OF THE THEORY OF THE EPR SPECTRA OF NITROXIDE RADICALS

The theory of the EPR spectra of stable nitroxide radicals has been presented in detail in a number of articles and monographs [1,5–7,21–24,

1

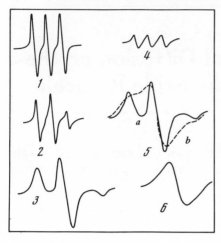

Figure 1. Electron paramagnetic resonance spectra of nitroxide radicals under model conditions: (*1*) alcoholic solution (10^{-3} M, 20°); (*2*) 80% glycerol solution in water (10^{-3} M, 20°); (*3*) 50% frozen glycerol solution in water (10^{-3} M, 77°K); (*4*) aqueous solution (10^{-3} M, 20°) in the presence of the paramagnetic ferricyanide ion (10^{-2} M); (*5*) water–glycerol matrix [(*a*) 10^{-3}; (*b*) 5×10^{-2} M; 77°K]; (*6*) aqueous solution (10^{-1} M, 20°).

35–42,52–57]. We shall limit ourselves here to the most essential information and special topics.

The paramagnetic properties of nitroxide radicals are due to the intrinsic angular momentum of the unpaired electron (spin). In a constant magnetic field of strength H_0, electron spins orient themselves in two directions, namely, along the field and against the field (the Zeeman effect).

Since these orientations interact differently with the magnetic field, two Zeeman energy levels with different populations arise in the system. Superposition of a perpendicular electromagnetic field of frequency ν results in reorientation of the spins and equalizing of the populations of the levels, if its energy ($h\nu$) equals the energy required for reorientation $g\beta H$, where h is Planck's constant, β is the Bohr magneton, and g is the g factor, which equals 2.0023 for a free electron. In real paramagnetic compounds the value of the g factor generally deviates from this value because of the small additional orbital magnetism arising as a result of the orbital motion of the electrons. Thus the condition for resonance is the equality

$$h\nu = g\beta H \tag{1.1}$$

The possibility of recording the absorption of electromagnetic radiation by paramagnetic groups in a constant magnetic field and the lineshapes in EPR spectra are largely determined by the processes involved in restoring the spin system to a state of thermal equilibrium (relaxation) following the primary absorption act. The most important characteristics of this process are the spin–spin relaxation time (T_2) and the spin–lattice relaxation time (T_1). Spin–lattice relaxation is due to the interaction between the electron spins and the magnetic and electric fields of the surroundings (lattice), which have fluctuating magnitudes and directions. This interaction results in a change in the populations of the Zeeman levels and, consequently, in the average energy of the spin system.

Fluctuating local magnetic fields, induced by the surrounding nuclear and electronic magnetic moments, are the cause of spin–spin relaxation. This type of relaxation causes a spread in the frequencies of the resonance transition between the Zeeman levels without altering the average energy of the spin system in these levels and is manifested in EPR spectra as broadening of the absorption lines. Although the nitroxide group arises as a result of homolytic cleavage of the O—H bond in the hydroxylamine derivative, the unpaired electron is delocalized over the oxygen and nitrogen. The unpaired electron in the spherically symmetrical S state undergoes a contact interaction with the spin of the nitrogen nucleus; this interaction causes splitting of the EPR lines into three components corresponding to orientation of the nuclear spin along the constant magnetic field ($M = +1$), perpendicular to the field ($M = 0$), and against the field ($M = -1$). Here M is the magnetic quantum number of the nucleus. This is the mechanism for the appearance of the so-called "isotropic hyperfine structure." Because of the spherical symmetry of the electron density in the S state, the result of this type of electron–nuclear interaction is independent of the angle of rotation between the radical and the magnetic field.

Another type of interaction is associated with configurations in which the unpaired electron is found in a p orbital, which does not possess spherical symmetry. Obviously, in this case the result will be dependent on the angle between the direction of the axis of the p orbital and the direction of the constant magnetic field along which the spins of the electron and the nitrogen nucleus are oriented. The additional local field induced by the nuclei will, as in the preceding case, alter the resonance conditions for an electron interacting with nuclei whose magnetic moment projections (M) equal $+1$, 0, and -1. The anisotropic hyperfine structure of EPR spectra arises in this manner. The rapid thermal motion of a radical, for example, in a medium of low viscosity, can average the anisotropic component of the hyperfine structure to zero.

The contribution of the electron–nuclear interaction to EPR linewidths is explained in a similar manner. Each specific case has its own set of local fields induced by the nuclei surrounding the unpaired electron because of the different orientations of the molecular axes of the radical relative to the direction of the magnetic field. Rapid thermal motion averages the anisotropic components of the magnetic fields and thereby narrows the EPR line. g-Factor anisotropy also has a significant effect on the type of EPR spectrum. The physical aspect of this effect is that the unpaired electron in the nitroxide group of the radical is under the influence of an asymmetrical electronic environment and that the small addition of orbital magnetism causing the g factor to deviate from 2.0023 will depend on the orientation of the radical and, therefore, on the intensity of its motion.

Thus the EPR spectra of frozen and supercooled solutions in the absence of rotational and translational motion are superpositions of "microspectra" of an enormous number of radicals with different orientations of the hyperfine and g-factor tensors relative to the external magnetic field. The motion of the radicals results in equalizing and averaging of the interactions, as well as simplification of the spectrum.

A far more complicated situation arises when nitroxide fragments are added to protein macromolecules, since the resultant motion of the nitroxide fragment depends on an extensive series of factors, namely, the rotational rates and mobilities of the macromolecule itself, the size and flexibility of the "leg" of the label, the microstructure of the protein, and the surrounding water in the vicinity of the labels added. It is wise to use simulation methods to analyze such systems. Theoretical and experimental analysis of various models of radical motion makes it possible to take into account the primary properties of a system and to disregard the secondary. Careful comparison of the quantitative experimental data with the results of theoretical calculations makes it possible, in the final analysis, to select the best model for the system under investigation.

THEORIES FOR THE EPR SPECTRA OF NITROXIDE RADICALS

The theory of the EPR spectra of stable nitroxide radicals in solution is based on the McConnell model [40]. In formulating the model it was assumed that the dependence of the shape of the EPR spectrum of nitroxide radicals on the rotational correlation time (τ_c) is effected through the influence of the anisotropy in the hyperfine and g-factor tensors on the lineshape parameters, particularly on the value of T_2.

The following are important questions in any theoretical treatment of the McConnell model:

1. In which range of correlation times (τ_c) does the motion occur?
2. Is the rotation isotropic or anisotropic?
3. What is the mechanism for the elementary motion of the radical?

It is first of all necessary to distinguish between rapid (with frequencies of 10^{10} to 5×10^8 sec^{-1}) and slow (5×10^8 to 10^7 sec^{-1}) motions. Different theories are employed to describe them. The theory of EPR spectrum linewidths and lineshapes in the rapid-motional region was developed in papers by McConnell [40], Kivelson [39], and Freed and Frenkel [57]. An analysis of the EPR spectra of nitroxide radicals in the slow-motional region is given in other works [52,53,59,63–65,78]. Very slow motions $(10^7$ to 10^5 sec$^{-1})$ were investigated in the work of Hyde [68].

Another important parameter in the theory of EPR spectra is the nature of the motion of the particles. The simplest motion is isotropic rotation, where the frequency of reorientation is the same in all directions. However, most paramagnetic probes are asymmetric particles, and their rotation is anisotropic. A theory that takes into account anisotropic rotation as applied to nitroxide radicals has been reported [58,60].

In the slow-motional region the mechanism of radical reorientation has a significant effect on the shape of EPR spectra [58,73–74]. The following three mechanisms are usually considered: (a) the Debye model of rotational Brownian motion, (b) the rotational random-jump model, which assumes instantaneous discontinuous reorientation of the particles alternating with a comparatively long state of rest, and (c) the model based on inertial effects, which assumes rapid changes in the orientation of a particle with gradual inertial slowing. The effectiveness of a theory is determined to a great extent by how well the models on which it is based correspond to the actual motion in the systems under investigation.

The following models are presently being used to analyze experimental data on nitroxide spin labels in biological systems: (a) isotropic rotation [39–41], (b) a "rotor on a rotor" [80], (c) anisotropic rotation in an isotropic medium [58,60], and (d) anisotropic motion in an anisotropic medium [54].

The first three models will be considered in the present chapter, and the fourth model, which is used chiefly to analyze data on the probing of membranes, will be described in Chapter 10.

THE ISOTROPIC ROTATION MODEL

Rapid-motional region. The McConnell model for the rapid-motional region was formulated mathematically in the work of Kivelson [39] and Freed and Frenkel [57]. The task of the theory was to find a quantitative relationship between the empirical parameter T_2, which is inversely proportional to the linewidth ΔH, and the microparameters of the radical: the components of the g factor (g_x, g_y, g_z) and hyperfine (A_x, A_y, A_z) tensors and the correlation time τ_c. Since both theories just mentioned start from similar assumptions and lead to similar results, we shall restrict ourselves here to a brief account of the main points of the work of Kivelson and McConnell [39–41].

Treatment of the system of energy levels for a nitroxide radical in a magnetic field involves analysis of the spin Hamiltonian [1,2]

$$H = \beta(g_x H_x \hat{S}_x + g_y H_y \hat{S}_y + g_z H_z \hat{S}_z) + h(A \hat{S}_z \hat{I}_z + B \hat{S}_x \hat{I}_x + C \hat{S}_y \hat{I}_y),$$

where S_x, S_y, and S_z and I_x, I_y, and I_z are the operators of the electron and nuclear spins corresponding to the different coordinate axes, g_x, g_y, and g_z are the g-factor components, and A, B, and C are the respective hyperfine interaction constants.

In the case of nitroxide radicals the theoretical treatment leads to the formula

$$\frac{T_2(0)}{T_2(M)} = 1 - (4/15 \cdot b \, \Delta\nu H \cdot T_2(0) \cdot M + \tfrac{1}{8} b^2 T_{2(0)} \cdot M^2) \tau_c, \quad (1.2)$$

which was derived on the basis of the following assumptions:

1. The motion of the radical is isotropic.
2. The magnetic anisotropy $(\Delta\nu)$ is small, that is,

$$\Delta\nu = \frac{|\beta|}{\hbar} [g_z - \tfrac{1}{2}(g_x + g_y)] \ll \nu;$$

3. None of the relaxation mechanisms except the electron–nuclear interaction (N_{14}) make anisotropic contributions.
4. The spectral lines are well resolved and do not overlap.
5. $(\pi A_{iso})^2 \tau_c^2 \ll 1$, where $|A_{iso}| = \tfrac{1}{3}|A + B + C|$ is the isotropic hyperfine interaction constant.
6. The anisotropy in the hyperfine structure is axially symmetric, that is, $C = A_{\parallel}$ and $A = B = A_{\perp}$.
7. $\omega^{-2} \ll \tau^2 \ll b^{-2}$, where $b = \tfrac{4}{3}\pi(A_{\parallel} - A_{\perp})$ and $\omega = g\beta H h^{-1}$.

Conditions 5 and 7 define the rapid-motional region. Equation (1.2) permits calculation of τ_c from the known parameters of the EPR spectra of radicals. For example, Stone et al. [41] use the following numerical values of the Hamiltonian parameters, which were determined experimentally for the di-*tert*-butylnitroxide radical:

$$A_{\parallel} = 87 \times 10^6 \text{ sec}^{-1}; \quad A_{\perp} = 14 \times 10^6 \text{ sec}^{-1};$$

$$g_x = 2.0089; \quad g_y = 2.0061; \quad g_z = 2.0027$$

In practice it is convenient to use equations obtained from (1.2) after substituting numerical values for the coefficients and taking into account the fact that the value of T_2 for each hyperfine-structure component is inversely proportional to the square of the intensity (height) of the EPR spectrum lines. We have, for example,

$$\nu_{(+1)} = 1/\tau_{c(+1)} = \frac{2 \cdot 10^8}{\left(\sqrt{\dfrac{h_0}{h_{(+1)}}} - 1\right) \Delta H_0} \text{ sec}^{-1} \quad (1.3)$$

and

$$\nu_{(-1)} = 1/\tau_{c(-1)} = \frac{3.6 \cdot 10^9}{\left(\sqrt{\dfrac{h_0}{h_{(-1)}}} - 1\right)\Delta H_0} \text{ sec}^{-1}, \tag{1.4}$$

where ΔH_0 is the width of the central component in gauss (G); h_0, $h_{(+1)}$, and $h_{(-1)}$ are the intensities of the spectral components with $M = 0, +1$, and -1 (Fig. 2); and $\nu = 1/\tau$ is a quantity called the "rotation frequency" of the radical by convention. We may also use the expression [57]

$$\nu_{(\pm1)} = \frac{1.2 \cdot 10^{10}}{\left(\sqrt{\dfrac{h_{(+1)}}{h_{(-1)}}} - 1\right)\Delta H_{(+1)}} \text{ sec}^{-1}. \tag{1.5}$$

The values of the g factors and hyperfine-structure parameters are only slightly dependent on the nature of the radicals. For example, the values of A_{iso} lie between 14.5 and 17 G for a large number of nitroxides with various structures. The solvent has a noticeable effect on the value of A_{iso}, varying it from 14.5 in nonpolar solvents to 16.7 G in water [1]. Usually fine differences in the values of the g and A tensors are not taken into account in approximate calculations, since other factors, particularly molecular motion, have a considerably stronger effect on the experimental parameters of the spectrum. However, these distinctions must be taken into account in more precise calculations. The "working" frequency range

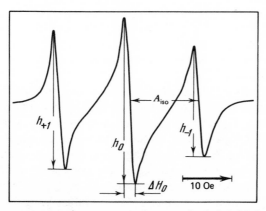

Figure 2. Electron paramagnetic resonance spectrum of spin label I (see p. 24) on the SH group of human serum albumin: h_0, h_{+1}, and h_{-1} are the intensities of the components with magnetic quantum numbers for the N^{14} nucleus (M) equal to 0, +1, and −1, respectively; ΔH_0 is the linewidth; and A_{iso} is the isotropic hyperfine interaction constant.

for Eqs. (1.3)–(1.5) is $5 \times 10^8 - 10^{10}$ sec^{-1}, the errors at the fringes of this range being 20–25%.

Several consequences of this theory have been confirmed empirically in experiments with solutions of radicals. For example, the dependences of the hyperfine linewidths on the viscosity [62] and on the frequency [61] of the EPR radiospectrometer fit the theoretical values.

Constant values for the coefficients b and $\Delta\nu$ for a given nitroxide radical are an experimental test for the applicability of the theory, since these quantities vary only as a result of chemical effects on the electronic structure. From Eq. (1.2) we can easily derive an expression that relates experimental values of T_2 to a set of "chemical" parameters of the radical

$$\delta = \frac{\Delta\nu H_0}{b} = \frac{(T_{2(0)}/T_{2(-1)}) - (T_{2(0)}/T_{2(1)})}{(T_{2(0)}/T_{2(-1)}) + (T_{2(0)}/T_{2(1)}) - 2}. \qquad (1.6)$$

Experiments have shown that in a number of systems (solutions, polymers, solutions of spin-labeled proteins, etc.) the changes in the parameter δ as the ambient conditions (temperature, pH, solvent, etc.) are varied greatly exceed the limits for the possible variations in $\Delta\nu$ and b. The most likely explanation for this effect is the assumption that the rotation of the radicals is anisotropic (see below).

When the rotational correlation time is increased ($\tau_c > 10^{-9}$ sec), as in the case of solutions of radicals in viscous liquids, the EPR spectrum changes qualitatively. The hyperfine-structure lineshape deviates significantly from the Lorentzian shape, and the value of A increases. Calculation of τ_c in these cases requires the special methods presented in the following section.

Slow-motional region. To calculate τ_c in the slow-motional region, the experimental spectra must be compared with spectra calculated with the aid of computers. The results of the calculations depend on the selection of the model of the motion.

Kuznetsov et al. [63] base their calculations of the spectra of a radical in a viscous isotropic medium on the assumption that changes in the orientation of the radical occur by means of random jumps. The time between two consecutive jumps is taken as the rotational correlation time. The calculated spectra (Fig. 3) are in good agreement with those obtained experimentally. In the rapid-motional region the correlation time is calculated directly from Eq. (1.5). The right-hand side of Eq. (1.5) is chosen as a convenient experimental parameter for τ_c values ranging from 10^{-9} to 5×10^{-9} sec. The relationship between this parameter, $\tau'_c = 1/\nu_{\pm1}$, and the corresponding τ_c values assigned in the case of the calculated spectra is illustrated in Fig. 4. Since for $\tau_c > 6 \times 10^{-9}$ sec the low-field extremum of the first-derivative spectrum is shifted, in the slow-motional region it is convenient to use the parameter κ:

$$\kappa = \frac{H(\tau) - H(\tau \rightarrow 0)}{H(\tau \rightarrow \infty) - H(\tau \rightarrow 0)} , \qquad (1.7)$$

where $H(\tau)$ is the position of the low-field maximum $(M = +1)$ of the first-derivative spectrum and $H(\tau \rightarrow \infty)$ and $H(\tau \rightarrow 0)$ are the positions of this maximum for the two limiting values of τ (see Fig. 5). In Kuznetsov's work [63] the value of the difference $\delta H = H(\tau \rightarrow \infty) - H(\tau \rightarrow 0)$ is taken equal to 10 Oe. A plot of the dependence of the parameter κ on τ_c is given

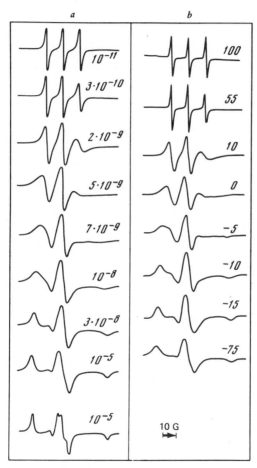

Figure 3. Theoretical (a) and experimental (b) EPR spectra of nitroxide radicals for different rotational correlation times (τ_c) (in sec) and different temperatures (in °C). The solvent is glycerol [63].

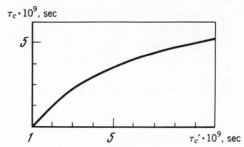

Figure 4.   Dependence of the rotational correlation time of a nitroxide radical (τ_c) on the parameter $\tau'_c = 1/\nu_{(\pm 1)}$, which is defined in Eq. (1.5) [63].

in Fig. 5. It is significant that κ is weakly dependent on the values of g and A and the linewidth. The main difficulty in using Eq. (1.7) in practice is obtaining the precise value of δH. The usual method for finding δH from the EPR spectra of frozen solutions may cause significant errors due to the difference between δH in the frozen solutions and under ordinary experimental conditions. These errors can be reduced to a minimum by selecting the best model medium.

The lineshapes in the spectra of nitroxide radicals in the 10^{-9}–10^{-7} sec correlation time range may be analyzed by a method that is based on the

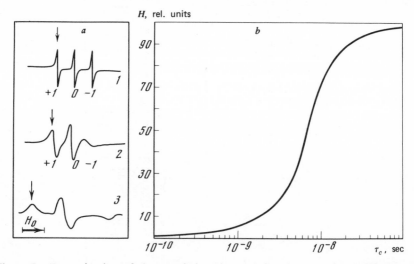

Figure 5.   Determination of the correlation time (τ_c) for slow rotational diffusion of nitroxide radicals from EPR spectral parameters [63]: (a) choice of parameters: (1) $H(\tau \to 0)$, (2) $H(\tau)$, (3) $H(\tau \to \infty)$; the arrows indicate the position of the parameters; (b) dependence of τ_c on the parameter κ [Eq. (1.7)].

use of the isotropic Brownian diffusion model [64,65]. For the calculation of τ_c, McCalley, Shimshick, and McConnell proposed using the shifts $\Delta H_1 = (H_{1(\infty)} - H_1)$ and $\Delta H_2 = (H_{2(\infty)} - H_2)$, where the indices 1 and 2 refer to the low-field and high-field components. The shift ΔH_2 turned out to be more sensitive to motion than ΔH_1. The value of $H_{(\infty)}$ is calculated from the experimental data by extrapolation to infinite viscosity. This method is applicable only in those cases in which the nitroxide fragment of the spin label is rigidly bound to a protein macromolecule. The method for finding τ_c is illustrated in Fig. 6.

A similar method was suggested by Goldman et al. [72]. Theoretical calculations of the spectra of nitroxide radicals at various assigned values of τ_c permitted Goldman et al. to derive the equation $\tau_c = a(1 - s)^{b_1}$, where the parameters a and b_1 depend on the individual linewidth, the rotation model, and the hyperfine-structure parameters, and s is the ratio between the outer peaks of the hyperfine structure and the corresponding values for solid matrices. The dependence of the values of the parameters on the rotation model is rather important. For example, in the $s \rightarrow 1$ region transition from the rotational Brownian motion model to the random-jump model decreases the calculated value of τ_c by almost an order of magnitude. This indicates that the choice of the best model is a very important factor in treating experimental data. According to the results obtained by Freed et al. [58], the model of decaying jumps should correspond better to the motion of low-molecular radicals, while in the

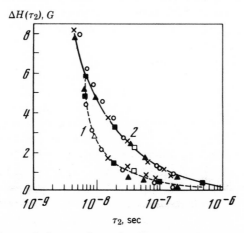

Figure 6. Dependence of the experimental shifts $\Delta H(\tau_2)$ in the EPR components of rotating nitroxide radicals on the correlation time (τ_2) [64,65]: (*1*) low-field component; (*2*) high-field component; the points correspond to different values for the elements of the *g* and *A* tensors.

case of labels rigidly bound to a macromolecule, the best fit with the experimental data should be expected from the Debye model of rotational Brownian motion.

In a number of papers [5,6,52,53] τ_c was evaluated in the slow-motional region by comparing the shape of the EPR spectrum of the radical in the system being investigated with its spectra in glycerol or a water–glycerol mixture at various temperatures. A variant of this approach was proposed [52,53]. Antsiferova, Lazarev, and Stryukov employed a model of random molecular reorientations in a viscous solution to calculate the spectra of nitroxide radicals with the ^{15}N isotope (the spin of the nucleus I equals $\frac{1}{2}$) as a function of correlation times ranging from 2.5×10^{-9} to 7.5×10^{-8} sec. The identification of the empirical spectra was aided by the Stokes–Einstein equation

$$\tau_c = \frac{4\pi a^3}{3kT} \eta, \qquad (1.8)$$

where η is the viscosity of the medium and a is the effective radius. After determining a from the experimental data for the rapid-motional region, then extrapolating them to more viscous media and performing parallel experiments with radicals enriched with ^{14}N and ^{15}N, these workers were able to evaluate τ_c by comparing the tagged spectra in the water–glycerol mixtures with the spectra in the systems under investigation.

The anisotropic nature of the rotations in the slow-motional region, as in the rapid-motional region, can cause differences between the theoretical and experimental spectra [72]. In particular, the EPR spectra of the peroxylamine disulfonate radical in 85% glycerol fit the theoretical spectra only after the anisotropic nature of the rotational diffusion of the radical has been taken into account.

The theory of the EPR spectra of radicals in systems with a nonuniform viscosity was developed by Freed and Frenkel [56]. It was assumed that both the rotational diffusion and the migration of radicals from one region to another are described with the aid of models based on random Markovian processes. For the sake of simplicity, the calculation was carried out for a radical with ^{15}N. As expected, the theoretical spectra of a two-phase system of radicals characterized by τ values equal to 10^{-10} and 10^{-9} sec were most sensitive to the transition frequency of the radicals (ν_{trans}) from one phase to the other, if $10^9 < \nu_{\text{trans}} < 10^{10}$ sec^{-1}. In this frequency range the spectral theory permits calculation of ν_{trans} from the parameters of the EPR spectra.

Ultraslow-motional region.   At correlation times ranging from 10^{-7} to 10^{-5} sec the methods presented above are not suitable because of the insensitivity of the EPR spectra to the frequency of spin reorientation. In this region we may use methods that are based on the sensitivity of the spin–lattice relaxation time T_1 to slow movements. These include electron–nuclear resonance and electron–electron double resonance (ELDOR) as well as the methods that exploit the phenomenon of saturation in EPR spectra.

Hyde [68] has shown that the dispersion signals obtained under the conditions of rapid adiabatic passage are sensitive to the rate of rotational diffusion of nitroxide radicals in cooled *sec*-butylbenzene. The saturation curves of the "usual" EPR spectra of the radicals also display a temperature dependence. The sensitivity of the parameters of the saturation curves to slow tumbling was also predicted [58].

The "rotor on a rotor" model (Wallach model) [80]. The basic principles of the Wallach model can be visualized from Fig. 7. The protein matrix undergoes Brownian rotation with a frequency of $\nu_p = \tau_p^{-1}$. The total mobility of a nitroxide fragment bound to a matrix by means of a "leg" depends on both ν_p and the flexibility and length of the "leg." In the simplest case the rotational motions around the joints connecting the individual links are divided by convention into two groups: (a) motions whose frequencies are considerably greater than the rotation frequency of the protein and (b) motions whose frequencies are considerably smaller than the rotation frequency of the protein. According to Wallach, the rotation frequency (ν) of a piperidine ring bound to a protein matrix is evaluated from the formula

$$\nu = \nu_p \cdot \alpha_1 \cdot \alpha_2 \ldots \alpha_n, \tag{1.9}$$

where $\alpha_1, \alpha_2, \ldots$ are factors that take into account the contribution of the internal rotations along the bonds of the "leg" joining the free-radical nitroxide group to the protein. Each internal rotation with a rate much greater than the rotation rate of the protein makes a contribution of $\alpha = \frac{1}{2}(3 \cos^2 \theta - 1)^2$, where θ is the angle between the axis being considered

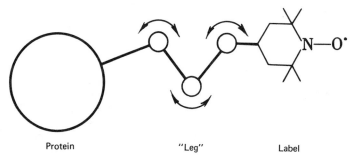

Protein   "Leg"   Label

Figure 7. The "rotor on a rotor" model according to Wallach [80].

and the axis of rotation of the adjacent bond; for bonds directly joined to the piperidine ring θ is the angle between the rotation axis and the z axis of the p orbital of the electron. The effect of rotations that are slower than the rotation of the protein is taken into account by means of an appropriate factor by averaging the possible orientations during rotation.

According to Wallach [80], α equals 10 for rapid rotations around C—C, C—O, and C—N bonds, α equals 4 for rotations around a bond with the ring, and α is approximately equal to 2 for very slow rotations in each bond. After determining the values of ν_p and ν experimentally, we can establish which rotational motions occur in each specific case. The confidence level of such an analysis is raised, if there are data for a whole set of labels with "legs" of different lengths and flexibilities.

The anisotropic rotation model. The effect of anisotropy in the rotational diffusion tensor of nitroxide radicals on their EPR spectrum parameters has been treated by Vasserman et al. [60]. It is assumed that the radical rotates like a gyroscope and that the rotation of the radical is effected by continuous diffusion.

The principal values of the diffusion tensor $D_1 = D_2 = D_\perp$ and $D_3 = D_\parallel$ are determined in the molecular coordinate system (Fig. 8). As in the case of the isotropic McConnell–Kivelson model, it is assumed that conditions 1–7 are fulfilled. In the present model the relaxational broadening of the spectral lines is determined chiefly by D_1.

The theory of EPR spectra relates the experimental parameter of anisotropy in spectra (ϵ) to the value of the anisotropy in the rotation of the

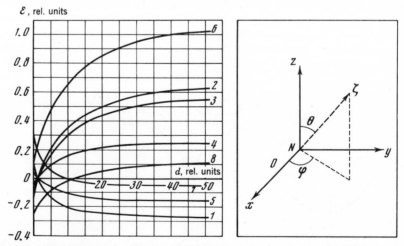

Figure 8. Dependence of ϵ on d for different orientations of the x, y, z, and ξ, θ, ϕ coordinate systems [60]. (See explanation in the text.)

radical d, which equals $D_\parallel/D_\perp$. The value of ϵ is calculated from the formula

$$\epsilon = \frac{T_{2(+1)}^{-1} - T_{2(0)}^{-1}}{T_{2(-1)}^{-1} - T_{2(0)}^{-1}} = \frac{\sqrt{(h_{(0)}/h_{(+1)})} - 1}{\sqrt{(h_{(0)}/h_{(-1)})} - 1}. \tag{1.10}$$

It is not difficult to prove [see Eqs. (1.2)–(1.3) and (1.10)] that

$$\left[\epsilon \sim \frac{\nu_{(+1)}}{\nu_{(-1)}}\right].$$

Calculation shows [60] that the type of dependence between ϵ and d is determined by two factors, namely: (a) the relative orientation of the system of principal axes for the g factor and hyperfine tensors and the principal axes for the rotational diffusion tensor and (b) the parameters describing the anisotropy in the g and A tensors.

Thus if the orientation of the rotation axes and the parameters describing the anisotropy are known, the theory of EPR spectra permits calculation of the value of d. Moreover, neglecting the rotational anisotropy (e.g., calculation according to $T_{2(-1)}$ and $T_{2(+1)}$ causes the order of the value of D_{iso} to be close to the value of $D_\perp$ ($0.2 \lesssim D_\perp/D_{\text{iso}} \lesssim 5$). This implies that the model of isotropic rotation generally describes the motion of the particles correctly. An exception is the case in which D_{iso} is calculated from Eq. (1.2) and the error due to neglect of the rotational anisotropy can result in a physical paradox, that is, negative values of ν. The calculation shows that if rotation with D_1 and D_2 occurs along the z axis, ϵ is independent of the rotation anisotropy ($\epsilon = 0$). The EPR spectra are most sensitive to anisotropy when the rotation occurs around the N→O bond (the x axis). The anisotropy in the EPR spectra (ϵ) is fairly strongly dependent on the parameters describing the anisotropy in the g and A tensors. In Fig. 8 curves 1–4 were calculated with averaged values of the g_i and A_i tensors, and φ and θ are polar angles with the following values: curve 1, 0 and $\pi/2$; curve 2, $\pi/2$ and $\pi/2$; curve 3, $\pi/3$ and $\pi/2$; curve 4, $\pi/2$ and $\pi/3$. Curves 5–8 were calculated with g_n^1 and a_n^1 values lying in the range of errors for the parameters that result in the following values for A_0, A_z, g_0, g_z, φ, and θ (rad); curve 5, 12.0, 14.6–0.25 $\times$ 10^{-3}, 2.5 $\times$ 10^{-3}, $\pi/2$, and $\pi/2$; curve 6, 12.0, -14.6, 2.9 $\times$ 10^{-3}, 1.2 $\times$ 10^{-3}, 0 and $\pi/2$; curve 7, -9.0, 11.2, 1.2 $\times$ 10^{-3}, -3.7×10^{-3}, $\pi/2$, and $\pi/2$; curve 8, 9.0, 11.2, 4.3 $\times$ 10^{-3}, 2.4 $\times$ 10^{-3}, 0, and $\pi/2$.

EXPERIMENTAL VERIFICATION OF THEORETICAL MODELS·

The experimental material available at the present time enables us to discuss the applicability of the various models of rotational diffusion to real spin-labeled proteins. At the same time, several additional approaches in-

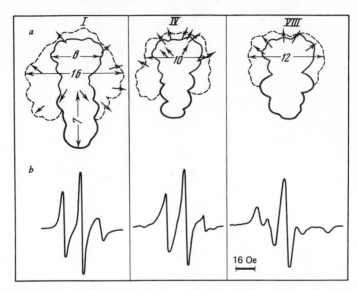

Figure 9. Projections of Stuart–Briegleb models (*a*) and the corresponding EPR spectra of paramagnetic labels (*b*). The solid thin lines represent the maximally stretched conformation, and the broken lines describe the contours of internal rotation around the single bonds. The arrows denote the possible projections of the N—Ó group. The solid thick lines characterize sections of labels I, IV, and VIII (see p. 24) joined to the β-93 SH groups of bovine hemoglobin and tightly compressed by the protein. The dimensions are given in Angström units, and the analysis is according to Wallach's method [80].

volving the use of labels of different lengths and flexibilities [3,25,32,81–87], measurement of the temperature dependence [25,81–87], and, finally, study of the accessibility of the nitroxide fragments of the labels to encounters with a spin probe [42,81,87] are important aids. These methods will be described in detail in the following chapters.

The data obtained with model systems are of great importance for a qualitative and semiquantitative interpretation of the spectra of labels on proteins. According to Figs. 1 and 9, different spectra of spin-labeled proteins contain elements that are characteristic of the spectra of nitroxide radicals dissolved in media of low viscosity, in glycerol solutions, in polymers, and even in glasses, in which the rotation of the radicals is practically stopped.

The isotropic rotation model.   Let us first consider the steric and conformational properties of the labels in use.

Let us assumé that the nitroxide group along with the ring is joined to a protein group by means of the respective "leg," whose position is fixed in a definite manner relative to the protein residue. We shall analyze the possible movements of the nitroxide group as a function of the length and flexibility of the "leg."

Different ways of varying the conformation of several labels are presented in Fig. 9. According to the figure, the free-radical end of long and flexible labels can assume various orientations despite the covalent bonding of the label to a protein group as long as there are no steric hindrances, and its motion actually imitates isotropic rotation, the rotational possibilities being somewhat more restricted in short labels.

Naturally, filling the space for free motion with bulky protein groups or water, whose viscosity is increased owing to the protein surface, should result in a limitation on the number of possible orientations for the nitroxide ring and slowing of certain motions.

The experimental data support this conclusion qualitatively. The motion of paramagnetic labels added to loose polypeptides and proteins whose secondary structure is disrupted in some manner is slowed by a factor of 10–40, in comparison to their free motion in an aqueous medium. The rotational frequency is correlated qualitatively with the distance from the polypeptide residue to the ring with the nitroxide group, increasing from 3.5×10^9 sec^{-1} to 4×10^9 sec^{-1} as the size of the label increases from 5 to 13 Å.

As will be shown in Chapter 7, the temperature dependences for the rotational diffusion of labels are substantially dependent on both the nature of the protein matrix and the structure of the labels themselves. Within the framework of the isotropic model this phenomenon can be described as a result of an effective change in the local viscosity of the medium in the vicinity of the labels added. This suggests that the isotropic model qualitatively describes real systems to some extent. However, in practice in every case there is some discrepancy between the experimental data and the quantitative laws that follow from this model. First of all, the equality between the values of $\nu_{(+1)}$, $\nu_{(-1)}$, and $\nu_{(\pm1)}$ predicted theoretically is not fulfilled. Furthermore, the experimental value of δ (Eq. (1.6)), which according to the theory can depend on the chemical nature of the radical and to a small extent on the nature of the solvent, does in fact differ significantly for different labels and protein matrices and varies with changes in the conditions, such as temperature.

The values of the effective energies and entropies of activation for rotational diffusion of labels calculated from the temperature dependence of $\nu_{(+1)}$ and $\nu_{(-1)}$ also differ substantially (see Chapter 7). All this forces us to turn our attention to other rotation models.

The Wallach model. The Wallach model relates the resultant motion of the nitroxide tip to the flexibility of the "leg" of the label and to the rotational diffusion of the protein macromolecule by means of Eq. (1.9), which is valid for the special case of maximally rapid rotation along the bonds of the "leg" of the label [80]. This equation can be rewritten in logarithmic form

$$\lg \nu = A + \lg \prod^{i} \alpha_i, \qquad (1.11)$$

where $\prod^{i} \alpha_i$ is the product of the Wallach coefficients for the bonds around which the maximally free or maximally slowed rotation is possible, beginning from any previously assigned section of the "leg."

We can present arguments supporting the fact that the Wallach model to some extent correctly reflects a number of fundamental properties of a label-protein macromolecule system. First of all, in almost all of the cases known the following law, which is predicted by Eq. (1.11), has been confirmed qualitatively, namely, that all other conditions being equal, the longer the label is, the larger is the resultant rotation frequency [3,32,81–84].

Furthermore, investigation of a large number of labeled proteins with paramagnetic probes (see Chapter 4) has shown that the nitroxide tips of labels on proteins extend outwardly from the protein surface into an aqueous environment whose viscosity generally approaches that of pure water. This implies that the main reason for the resultant slowing of the labels on proteins (by a factor of $\leq 10^3$) is that the protein matrix more strongly clutches some of the labels closer to the protein surface. This idea clearly corresponds qualitatively to the Wallach model.

Equation (1.9) has been tested quantitatively on sperm-whale myoglobin with labeled histidine groups, on hemoglobin labeled at the β-93 SH groups, and on lysozyme with a modified histidine-15 group. As Fig. 10 indicates, a satisfactory correlation was observed between log ν and log $\prod^{i} \alpha_i$. However, the fit between the limiting versions of the theory and the experiment was only qualitative. The slopes in the respective coordinates equal 0.33 for myoglobin, 0.8 for hemoglobin, and 0.8 for lysozyme. On the basis of the discrete version of the Wallach theory, which divides all the bonds into two categories, that is, rapidly and slowly rotating, it is difficult to interpret the data on the temperature dependence of the EPR spectra of labels on proteins.

The dependence of the parameter ν on the viscosity for spin-labeled proteins similarly does not correspond quantitatively to the simplified version of the Wallach theory [81,83,88].

The model of anisotropic rotation in an isotropic medium. As we saw in the preceding sections, the isotropic rotation and "rotor on a rotor" models are very rough approximations of the behavior of the nitroxide fragments of labels on proteins, and Eqs. (1.3), (1.5), and (1.9) do not

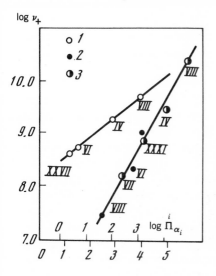

Figure 10. Dependence of the experimental parameters $\log \nu^+$ on the theoretical values of $\log \Pi_{\alpha_i}^i$. Here ν is the rotation frequency of the spin labels, and $\Pi_{\alpha_i}^i$ is Wallach's parameter, which characterizes the length and flexibility of the labels [81–83]: (*1*) myoglobin histidine groups (lower scale on the abscissa); (*2*) hemoglobin β-93 SH group; (*3*) lysozyme (upper scale on the abscissa).

correspond to the experimental data on spin-labeled proteins. We can formulate additional laws that are observed experimentally, but cannot be understood with the aid of the foregoing models:

1. The values of the effective energies and entropies of activation calculated from the temperature dependence of $\nu_{(-1)}$ and $\nu_{(+1)}$ differ from each other in such a way that $E^+ \gtrsim E^-$ and $\Delta S^+ \gtrsim \Delta S^-$.
2. The values of $\nu_{(\pm 1)}$ calculated from the "isotropic" formula (1.5) are close to $\nu_{(-1)}$ but significantly different from $\nu_{(+1)}$.
3. When the structure (length and flexibility) of the spin labels and the protein structure are varied, and when the ambient conditions (pH, temperature, composition of the medium, etc.) are altered, the most sensitive parameters are $\nu_{(+1)}$, E^+, and ΔS^+. The quantities E^- and ΔS^- are, in comparison, only slightly dependent on all of these conditions.

All of these facts are readily explained by adopting the anisotropic rotation model [60], that is, by assuming that the motion of the nitroxide fragment of the labels is similar to that of a gyroscope.

In fact, if we follow the experimental data obtained according to the spin probe–spin label method (see Chapter 4) and assume that the nitroxide fragment of the labels is pointed outward from the protein surface, we may offer the following physical picture for the motion of the nitroxide fragment. The piperidine ring of the radical rotates very rapidly ($D_1 \sim D_2$) around the bond directly adjoining the ring. The angle between this bond and the N→O axis (x axis) is 109° (Fig. 8). Inasmuch as the motion around the other bonds may somewhat alter the nature of the motion of the nitroxide fragment, we may assume in an approximate treatment that the value of the angle between the resultant rotation axis and the x axis is $90°$ and that the rotation occurs around the y axis ($\varphi = \pi/2$, $\theta = \pi/2$). The motion around the other axes results from rotational movements around the joints of the "leg." These movements are slower due to the hindering effects of the protein matrix. It is not difficult to see that this picture is consistent with the motion of an anisotropic gyroscope that rotates rapidly around the y axis ($D_1 = D_2$) and slowly around the other axes.

This model accounts well for the differences between the experimental values of $\nu_{(+1)}$ and $\nu_{(-1)}$ as well as between E^+ and E^-, the similarity between $\nu_{(\pm 1)}$ and $\nu_{(-1)}$ as well as between E^+ and E^-, the positive value of E, and the sensitivity of E^+, ΔS^+, and $\nu_{(+1)}$ to changes in the state of the protein matrix and to variations in the structure of the "leg" of the radical. This is due to the fact that the parameters $\nu_{(-1)}$ and E^- are related to the motion (D_1) of the nitroxide ends of the labels, which are far from the protein matrix and thus have motion that is practically insensitive to its structure. In addition, $\nu_{(+1)}$ and E^+ are related, in the final analysis, to the motions of the "tails" of the labels, which come directly in contact with the protein matrix and react with considerably more sensitivity to changes in its state.

We may, therefore, state that there is better qualitative agreement between the anisotropic rotation model and the experimental data on spin labels attached to proteins than in the case of the other available models. We shall, therefore, use this model for a semiquantitative analysis of the experimental data without ignoring its limitations: (a) the uncertainty in the choice of the exact directions of the rotation axes relative to the molecular axes, which is liable to vary with the experimental conditions (viscosity, temperature, etc.) and (b) the lack of precise values for the parameters describing the anisotropy in the g factor and hyperfine structure for radicals in polar media (the data presently available permit selection of the most probable values for these quantities) and the possibility that they vary under different conditions.

Moreover, these limitations are inherent to some extent even in simpler systems such as solutions of radicals in liquids and polymers. However, the results reported by Vasserman et al. [60] indicate that the anisotropic diffusion model satisfactorily describes the behavior of radicals with different structures in liquid and polymeric media.

The foregoing implies that the use of the anisotropic rotation model permits a more quantitative interpretation of the data on the rotational diffusion of spin labels attached to proteins.

There is every basis to assume that further improvement in the methods for calculating EPR spectra with the aid of computers will make possible a rigorously quantitative description of the behavior of spin labels in biological systems in the framework of the anisotropic diffusion and the "rotor on a rotor" models. Progress in this direction will undoubtedly advance electron–nuclear resonance and ELDOR, as well as the techniques based on rapid adiabatic passage and a combined analysis of the spectra of radicals enriched with ^{14}N and ^{15}N isotopes [77].

Chapter Two
Synthesis of Spin-labeled
Compounds: Spin Labels

The actual usefulness of the spin-label technique is greatly dependent on the possibility of the specific addition of nitroxide radicals to certain groups in a protein. Therefore, many of the labels used contain active groups that are fragments of known reagents whose selectivity has been verified by a great deal of experimental material.

At the present time there is a wide assortment of spin labels that react more or less specifically with the functional groups of proteins. Each label may be divided by convention into three main parts: (a) a paramagnetic fragment (a pyrrolidine or piperidine nitroxide ring or a transition-metal complex), which provides an intense EPR signal, (b) a chemically active residue, which is used for covalent or coordinate bonding to a specific protein group, and (c) a linking group of atoms (the "leg"), whose structure (length, flexibility, hydrophobic nature, and polarity) largely determines the behavior of the label following addition to a protein matrix.

The paramagnetic fragment is most often a nitroxide radical of the pyrrolidine or piperidine series. As already noted, these compounds are distinguished by the high chemical stability of the nitroxide group [35]. These groups are very resistant to oxidation by oxygen and other oxidizing agents. In neutral and alkaline media nitroxide radicals are scarcely reduced by protein groups in dilute solutions. These radicals are reduced efficiently in the presence of proteins only in acidic media at pH < 3. In the presence of oxidation–reduction biological systems the radicals may be reduced with great intensity, causing the EPR signal to vanish.

A convenient starting material for obtaining spin labels in reactions that do not affect the free valence is the free nitroxide radical 2,2,6,6-tetramethyl-4-oxopiperidine-1-oxyl (R):

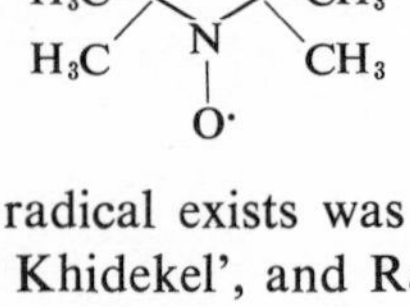

The hypothesis that this radical exists was first made according to its EPR spectrum by Lebedev, Khidekel', and Razuvaev [89].

Radical R was first obtained in pure form from triacetonimine in the label-atoms laboratory of the Institute of Chemical Physics of the USSR Academy of Sciences by Rozantsev and Mamedova [35]. The original triacetonimine can be obtained by two methods, either by starting from phorone and ammonia or by direct condensation of three molecules of acetone with one molecule of ammonia [36].

These workers, together with Neiman, were more recently the first to carry out a whole series of nitroxide radical reactions that do not affect the free valence [36]. This type of reaction presently underlies practically all the syntheses of spin labels.

A list of the chemical formulas for the spin labels and groups modified is given below. According to this list (p. 24), spin labels generally include fragments of already known, widely accepted specific reagents for protein residues or prosthetic groups. Figure 11 provides a visual representation of the relative dimensions of the nitroxide tips and the chemically active legs of several labels.

Further information on the methods of synthesis and the properties of nitroxide radicals can be found in various works [7,36,90]. The syntheses of the labels have been described in the references cited in the list on p. 24.

We use the following terms. A label is a compound covalently bond to functional groups of biological systems. A probe is a substance whose

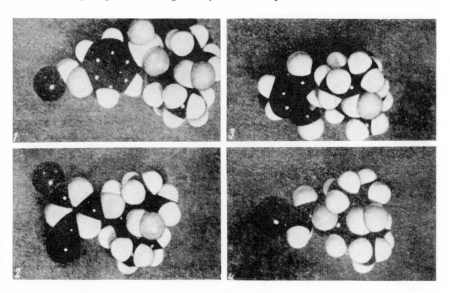

Figure 11. Stuart–Briegleb models for several spin labels: (*1*) XXVII; (*2*) VII; (*3*) VIII; (*4*) IV.

List of Chemical Formulas for Spin Labels and Protein Groups Modified

Amino Acid, Protein Group Modified	Formula of Spin Label		Reference
Cysteine, SH	Cl—Hg—⟨benzene ring⟩—$\overset{\displaystyle C}{\underset{\displaystyle O}{\|}}$—$\overset{\displaystyle N}{\underset{\displaystyle H}{\|}}$—R$_6$*	(I)	[20, 25, 31, 92, 94]
Same	Cl—Hg—⟨benzene ring⟩—$\overset{\displaystyle C}{\underset{\displaystyle O}{\|}}$—$\overset{\displaystyle N}{\underset{\displaystyle H}{\|}}$—R$_5$**	(II)	[95a]
"	Cl—Hg—⟨benzene ring⟩—$\overset{\displaystyle C}{\underset{\displaystyle O}{\|}}$—O—R$_6$	(III)	[96]
Cysteine, SH; histidine, imidazole	I—CH$_2$—$\overset{\displaystyle C}{\underset{\displaystyle O}{\|}}$—$\overset{\displaystyle N}{\underset{\displaystyle H}{\|}}$—R$_6$	(IV)	[1—3, 28, 31, 32, 97, 97a, 98]
Same	I—CH$_2$—$\overset{\displaystyle C}{\underset{\displaystyle O}{\|}}$—$\overset{\displaystyle N}{\underset{\displaystyle H}{\|}}$—R$_5$	(V)	[97]
"	I—CH$_2$—$\overset{\displaystyle C}{\underset{\displaystyle O}{\|}}$—O—R$_6$	(VI)	[98]
Lysine, histidine, imidazole	chloro-triazine ring: Cl—⟨triazine⟩—$\overset{\displaystyle H}{\underset{}{N}}$—R$_6$ (with Cl substituent)	(VII)	[30, 32, 92, 99, 100]
Cysteine, SH; lysine, ε-NH$_2$	maleimide ring: HC═CH, C=O, N—R$_6$ or R$_5$, C=O	(VIII)	[1, 32, 33, 42, 95, 102]
Same	maleimide: HC═CH, C=O, N(—CH$_2$)$_n$—$\overset{\displaystyle N}{\underset{\displaystyle H}{\|}}$—$\overset{\displaystyle C}{\underset{\displaystyle O}{\|}}$—R$_5$, C=O $n=1,2,3,8$	(IX)	[102]

Amino Acid, Protein Group Modified	Formula of Spin Label	Reference

Cysteine, SH; lysine, ε-NH$_2$

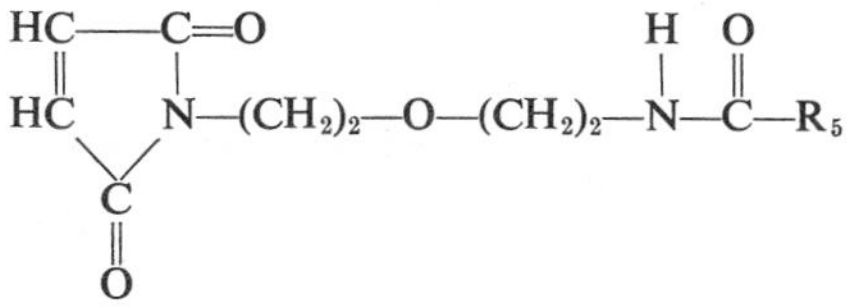

$$\text{HC---C=O} \qquad \qquad \text{H} \quad \text{O}$$
$$\text{HC} \quad \text{N---(CH}_2)_2\text{---O---(CH}_2)_2\text{---N---C---R}_5 \quad \text{(X)} \quad [102]$$
$$\text{C}$$
$$\text{O}$$

Cysteine, SH

$$\text{O}$$
$$\text{I---CH}_2\text{---C---(CH}_2)_n\text{---C---CH}_3 \qquad \text{(XI)} \quad [103]$$
$$n = 3,4,5,6 \quad \cdot\text{O---N} \quad \text{O}$$
$$\text{H}_3\text{C---C------CH}_2$$
$$\text{CH}_3$$

Serine, OH (in active centers of enzymes)

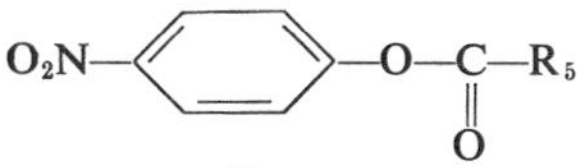

$$\text{O}_2\text{N---⟨⟩---O---C---R}_5 \qquad \text{(XII)} \quad [65, 104]$$
$$\text{O}$$

Same

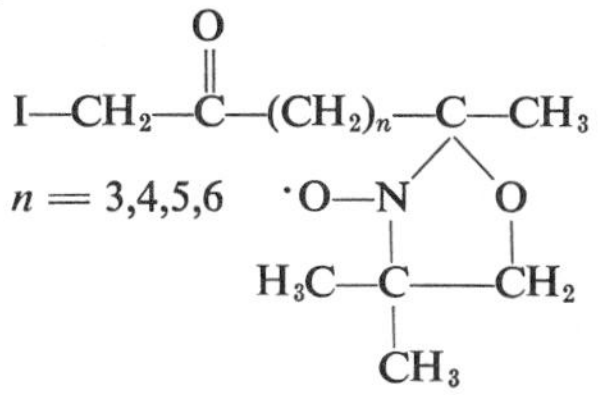

$$\text{O}$$
$$\text{F---P---O---R}_6 \qquad \text{(XIII)} \quad [105]$$
$$\text{O} \qquad \text{R---cyclohexyl}$$
$$\text{R}$$

"

$$\text{O}$$
$$\text{F---P---O---R}_6 \qquad \text{(XIV)} \quad [106]$$
$$\text{O---R}$$

"

$$\text{O}$$
$$\text{F---P---O---R}_6 \qquad \text{(XV)} \quad [106]$$
$$\text{CH}_3$$

"

$$\text{O}$$
$$\text{O}_2\text{N---⟨⟩---O---P---R}_6 \qquad \text{(XVI)} \quad [107, 108]$$
$$\text{CH}_3$$

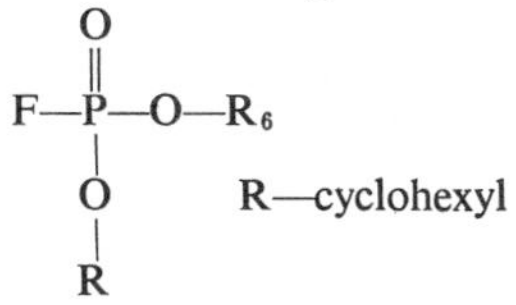

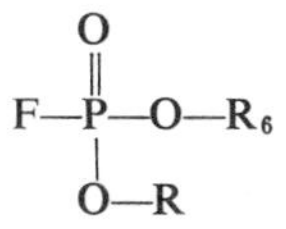

"

$$\text{O}_2\text{N---⟨⟩---O---C---(CH}_2)_n\text{---C---CH}_3$$
$$\text{O} \quad \cdot\text{O---N} \quad \text{O} \quad \text{(XVII)} \quad [103]$$
$$\text{H}_3\text{C---}|$$
$$\text{CH}_3$$

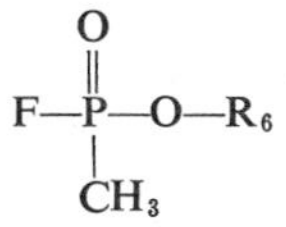

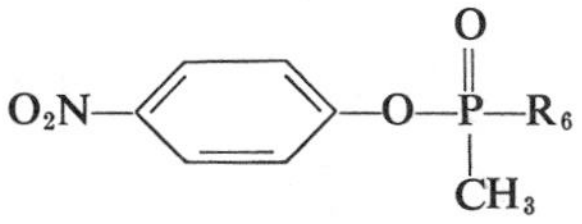

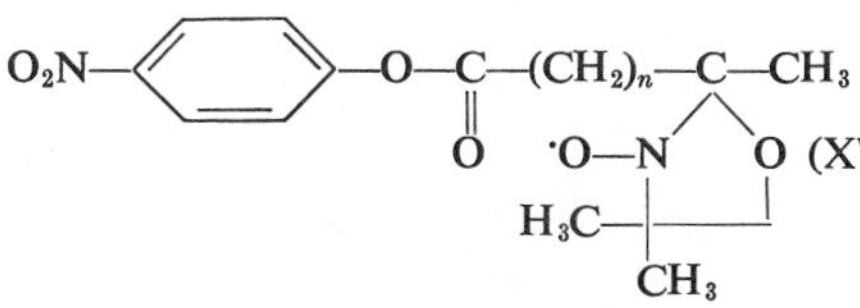

25

Amino Acid, Protein Group Modified	Formula of Spin Label	Reference
Serine, OH (in active centers of enzymes)	O_2N-⟨benzene ring with NO_2⟩$-O-\underset{\underset{O}{\|\|}}{C}-(CH_2)_2-\underset{\underset{O}{\|\|}}{C}-\underset{\underset{H}{\|}}{N}-R_6$ (XVIII)	[109]
Same	O_2N-⟨benzene ring⟩$-O-\underset{\underset{O}{\|\|}}{C}-(CH_2)_2-\underset{\underset{O}{\|\|}}{C}-O-R_6$ (XIX)	[109]
"	O_2N-⟨benzene ring⟩$-O-\underset{\underset{O}{\|\|}}{C}-(CH)_2-\underset{\underset{O}{\|\|}}{C}-\underset{\underset{H}{\|}}{N}-R_6$ (XX)	[109]
Serine-195, OH (in α-chymotrypsin)	⟨steroid-like structure with nitroxide⟩$-O-\underset{\underset{CH_3}{\|}}{\overset{\overset{O}{\|\|}}{P}}-F$ (XXI)	[65]
Histidine, imidazole	$Br-CH_2-\underset{\underset{O}{\|\|}}{\overset{}{C}}-\underset{\underset{}{\|}}{\overset{\overset{H}{\|}}{N}}-R_5$ (XXII)	[110]
Same	$Br-\underset{\underset{CH_2-COOH}{\|}}{\overset{\overset{H}{\|}}{C}}-\underset{\underset{}{}}{\overset{\overset{O}{\|\|}}{C}}-\underset{}{\overset{\overset{H}{\|}}{N}}-R_5$ (XXIII)	[110]
"	$Br-\underset{\underset{COOH}{\|}}{\overset{\overset{H}{\|}}{C}}-CH_2-\underset{\underset{O}{\|\|}}{C}-\underset{}{\overset{\overset{H}{\|}}{N}}-R_5$ (XXIV)	[110]
"	$Cl-CH_2-\underset{\underset{CH_2-\text{⟨ring⟩}}{\|}}{\overset{\overset{O}{\|\|}}{C}}-\underset{}{\overset{\overset{H}{\|}}{C}}-\underset{}{\overset{\overset{H}{\|}}{N}}-R_6$ (XXV)	[111]
Histidine; imidazole; lysine, ε-NH$_2$	⟨triazine ring with Cl, Cl⟩$-x(Cu^{2+})$ (Procion 2RP) (XXVI)	[3, 19, 20, 81, 92]

Amino Acid, Protein Group Modified	Formula of Spin Label		Reference
Histidine; imidazole; lysine ε-NH$_2$	Cl—N$_2$—⟨benzene⟩—S(=O)(=O)—O—R$_6$	(XXVII)	[92]
Same	O=C=N—R$_6$	(XXVIII)	[41]
Lysine, ε-NH$_2$; terminal NH$_2$	(succinimide ring) N—O—C(=O)—CH$_2$—R$_6$	(XXIX)	[99]
Methionine, S-CH$_3$ (in acid pH)	Br—CH$_2$—C(=O)—N(H)—R$_6$	(XXX)	[111]
Asparagine, carboxyl	N$_2$—CH—C(=O)—R$_5$	(XXXI)	[113, 114]
Lysine, ε-NH$_2$	O=⟨piperidine⟩N—O·	(XXXII)	[115]
Hydrophobic pocket of the active center of α-chymotrypsin	O$_2$N—⟨indoline (C$_2$H$_5$)$_4$⟩N—O·	(XXXIII)	[27]
Same	C$_6$H$_5$—C(H)—C(=O)—N(H)—R$_6$ with Br	(XXXIV)	[116]
Active center of dehydrogenase	H—C(=O)—(CH$_2$)$_n$—C(CH$_3$)⟨O·—N O⟩	(XXXV)	[103]
Same	(adenine NH$_2$ ... ribose HO, OH)—CH$_2$—O—P(=O)(O)—O—P(=O)(OH)—O—R$_6$	(XXXVI)	[117—119]

Amino Acid, Protein Group Modified	Formula of Spin Label		Reference
Active center of dehydrogenase		(XXXVII)	[120]
Active center of actin		(XXXVIII)	[121]
Active center of myosin	$ATP-R_6$	(XXXIX)	[122]
Active center of lysozyme		(XL)	[123, 124]
Active center of carbonic anhydrase		(XLI)	[125]
Same	Sulfonamide derivatives	(XLII)	[126]
Hydrophobic portion of serum albumin		(XLIII)	[44, 127]
Active center of ribonuclease		(XLIV)	[128]

List of Chemical Formulas for Spin Labels and Protein Groups Modified (*Continued*)

Amino Acid, Protein Group Modified	Formula of Spin Label	Reference
Regulator center of hemoglobin	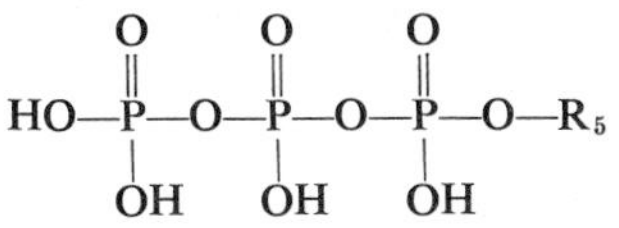 $$HO-\overset{\overset{O}{\|\|}}{\underset{\underset{OH}{\|}}{P}}-O-\overset{\overset{O}{\|\|}}{\underset{\underset{OH}{\|}}{P}}-O-\overset{\overset{O}{\|\|}}{\underset{\underset{OH}{\|}}{P}}-O-R_5 \quad (XLV)$$	[128, 134]
Same	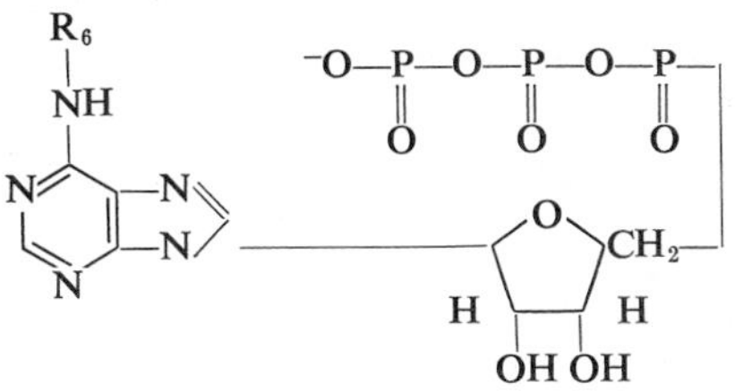 (XLVI)	[129, 134]
Heme Groups	$N\equiv C-R_6 \quad (XLVII)$	[130—133]
Nonheme iron-containing center of nitrogenase	$N\equiv C-R_5 \quad (XLVIII)$	[135]
Active centers of antibodies	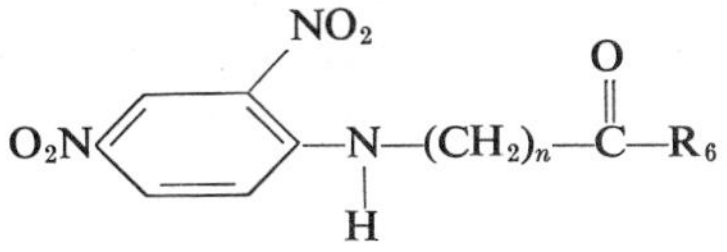 $O_2N\langle\bigcirc\rangle-\overset{\overset{NO_2}{\|}}{\underset{\underset{H}{\|}}{N}}-(CH_2)_n-\overset{\overset{O}{\|\|}}{C}-R_6 \quad (XLIX)$	[136]
Serine-195, OH (α-chymotrypsin)	*Ortho, meta* and *para* derivatives of benzyl sulfonyl fluoride (L)	[137]
Tyrosine, OH	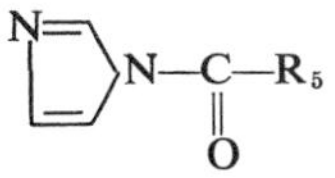 $N-\overset{\|}{\underset{\underset{O}{\|\|}}{C}}-R_5 \quad (LI)$	[112]

** R_5

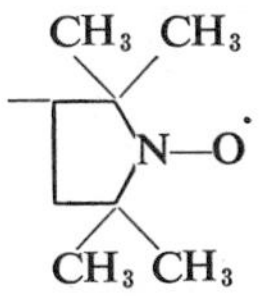

$$\begin{array}{c} CH_3 \quad CH_3 \\ \diagdown \diagup \\ \boxed{} N-O^{\cdot} \\ \diagup \diagdown \\ CH_3 \quad CH_3 \end{array}$$

* R_6:

$$\begin{array}{c} CH_3 \quad CH_3 \\ \diagdown \diagup \\ \hexagon N-O^{\cdot} \\ \diagup \diagdown \\ CH_3 \quad CH_3 \end{array}$$

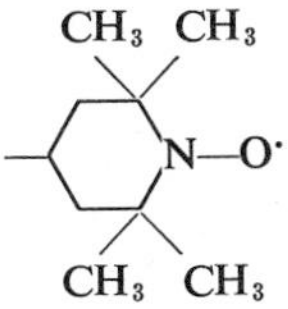

addition to the matrix under investigation is effected by noncovalent inter-actions. The term "nitroxide radical" is employed in this monograph. Other terms are also encountered in the literature [36].

CHEMICAL MODIFICATION OF PROTEINS BY SPIN LABELS

The main principles of introducing spin labels into proteins are almost the same as the principles of ordinary chemical modification. However, the use of spin-labeled compounds often greatly facilitates the control over the course of the modification, since the EPR spectra of nitroxide radicals are significantly altered following addition of the radicals to proteins (see Fig. 9). An important condition for the application of the spin-label tech-nique is the implantation of the labels on specific protein groups.

The synthesis of spin-labeled protein preparations is carried out in several stages:

1. Incubation of the protein solution with a solution of the label. Excess concentrations of the spin labels are usually employed. However, where the chemically active portion is highly specific, an equivalent amount of the reagent is added. This makes it possible to directly follow the kinetics of the modification by EPR.
2. Purification of the spin-labeled preparation from traces of the unreacted reagent. Dialysis and chromatography on Sephadex are the most con-venient purification methods. When several modification products are obtained, it is advisable to separate them.
3. Determination of the number of labels (n) added to the protein. The value of n can be determined to within 5% by quantitatively comparing the EPR spectra of frozen solutions of labeled proteins and dilute standard solutions of the radicals in water–glycerol glasses. In either case the EPR spectra generally correspond to individual unreacted radi-cals and have practically the same lineshape. This permits calculation of the concentration of the radicals from the ratio between the heights of the standard and experimental signals. In the case of distortions in the lineshape, as in the presence of paramagnetic groups or high ni-troxide concentrations ($[C] > 10^{-2}$ M), it is necessary to integrate the spectra, which may reduce the accuracy of the determination of n up to 25%.

IDENTIFICATION OF THE SITE OF
THE ADDITION OF LABELS TO
PROTEIN MACROMOLECULES

The principles of identification are generally analogous to those used for ordinary chemical modification. The identification process includes such operations as carrying out specific analytical reactions for certain protein groups, preliminary blocking of the groups modified, amino-acid analysis, composing peptide maps, and so forth. However, the addition of spin labels to biological macromolecules opens up new possibilities that facilitate identification. As will be shown in Chapters 3 and 5, in many cases EPR and NMR permit evaluation of the distance between labels and between labels and specific protons. When there is a three-dimensional X-ray diffraction analysis model of the protein, this aids in making a choice among alternative possibilities of label addition.

CONTROL OVER THE BIOLOGICAL ACTIVITY AND MACROMOLECULAR STRUCTURE OF SPIN-LABELED PROTEINS

The nitroxide fragments of most spin labels contain both a polar N—O˙ group and a hydrophobic aliphatic portion, which make it a peculiar hydrophobic–hydrophilic reagent capable of reacting moderately with both polar and nonpolar groups.

The combination of these properties provides labels with a fair degree of solubility in aqueous solutions and, at the same time, causes a very weak physical interaction between the nitroxide fragments and the proteins. Even when the component concentrations are very high, chemically inert five- and six-membered nitroxides generally do not have an effect on the enzymatic activity and physicochemical properties of proteins. Thus except in the special cases in which hydrophobic aromatic spin probes are employed, the chemical reaction and the interaction between the legs of the labels and the protein groups are responsible for the main result of the action of labels on proteins. The general tendency is that modification of the surface groups of a protein that are not part of the active center has practically no effect on the biological activity or the overall macromolecular structure of the protein. Interactions with functional residues more deeply embedded in the protein matrix somewhat disrupt the local structure of proteins in the vicinity of the addition. Specific examples of the modification of proteins and the identification of the site of the addition of labels to macromolecules will be given in the following.

REACTION OF SPIN LABELS WITH PROTEIN SULFHYDRYL GROUPS

Selective modification of protein sulfhydryl groups is greatly facilitated in the presence of very specific reagents, including mercurials and alkylating reagents [1,3,25,92,95,138]. The reaction between horse hemoglobin sulfhydryl groups and a reaction mixture containing the nitroxide derivative of *p*-chloromercuribenzoate (PCMB-radical II) has been described [95]. Experiments with purified preparations of radicals I and II in bovine hemoglobin solutions have confirmed the main conclusions drawn by Griffith and McConnell [95]. After incubation of a bovine hemoglobin solution with label I at pH 7 in a 0.05 M phosphate buffer for 12–15 hr at a temperature of 2–3° C, the mixture was subjected to gel filtration through Sephadex G-50. Spectrophotometric analysis of the fractions for hemoglobin content and EPR analysis for radical content revealed that the fractions containing hemoglobin produce EPR signals indicative of the presence of an immobilized radical (see Fig. 9). The intensity of the EPR signal was proportional to the protein concentration. Prior blocking of the SH groups with PCMB prevented implantation of the label. The spectrum of the radical in the bovine hemoglobin was analogous to the corresponding spectrum of the radical in horse hemoglobin labeled by a nitroxide group at the β-93 SH group. In those cases in which the ratio between radical I and hemoglobin in the incubation mixture was 1:1, there was practically quantitative addition of the radical, the kinetics of which could be followed by EPR. The radical added is tightly held by the protein during chromatography on Sephadex and as the pH of the solution is varied from 4 to 11.

The reaction, purification, and identification of the reaction products of hemoglobin with the alkylating radicals IV, V, and VII are carried out in a similar manner. In all cases prior blocking of the sulfhydryl groups prevents addition of the labels [97].

The precision of the implantation of spin labels IV and VIII on the hemoglobin β-93 sulfhydryl group has been confirmed unambiguously by X-ray diffraction analysis [193]. This method also made it possible to establish in detail the effect of dense packing of the label on the protein structure. It turned out that a nitroxide label fragment entering the hydrophobic pocket of the macromolecule moves the protein groups of different subunits somewhat apart, resulting in a decrease in the degree of cooperativity in the hemoglobin subunit system.

Investigations of the reactions of a whole series of spin labels that are specific reagents for sulfhydryl groups with aspartate aminotransferase (AAT) revealed differences between their reactivities [33]. The mercurials, as expected, proved to be the most active and thoroughly acting reagents. The use of peptide maps (Fig. 12) confirmed the covalent addition of the labels to the sulfhydryl groups of the enzyme.

Reactions of spin labels for histidine groups. A whole series of alkylating reagents for histidine groups in proteins has been described in the literature. Radicals VII and XXII, whose chemical structure suggested that

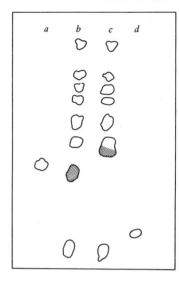

Figure 12. Distribution of sulfur-containing peptides from a chymotrypsin hydrolysate on a chromatogram [33]: (*a*) methionine (marker); (*b*) hydrolysate of original carboxymethylated enzyme; (*c*) same following prior modification by radical VI; (*d*) dye.

they would be added covalently to histidine imidazole rings, have been synthesized.

In particular, it has been established [12,30–33,92] that compounds containing the dichlorotriazine residue can be added to deprotonated forms of protein histidine and lysine groups by a nucleophilic substitution reaction that results in the evolution of HCl and the formation of a strong covalent bond. This reaction involves only deprotonated forms of imidazole and ε-amino groups. When lysozyme is reacted with the trichloroazinyl derivative of nitroxide radical VII at pH 7, only one molecule of the radical is added. After blocking the only histidine residue in lysozyme by the specific reagent 5-amino-I-H-tetrazolium monohydrate according to the technique outlined by Likhtenshtein and Akhmedov [30], the addition of radical VII to the lysozyme molecule does not occur.

Additional arguments in favor of the reaction between radical VII and histidine groups have been obtained with albumins and hemoglobin [92]. Incubation of a solution of bovine serum albumin, whose surface has only 10 easily accessible histidine groups at room temperature and neutral pH, with radical VII yielded spin-labeled albumin preparations. However, under similar conditions bovine hemoglobin, which does not possess reactive histidine groups at neutral pH, does not take on any appreciable number of radicals [92]. It appears that radical VII is capable of reacting with easily accessible histidine groups. A similar result was obtained with sperm-whale myoglobin. The easily accessible surface histidine groups of sperm-whale myoglobin are readily modified by label VII [32,81,83]. Spin reagents based on α-haloacyl radicals (e.g., derivatives of iodo- and bromoacetates and acetamides) are also used to modify comparatively reactive histidine residues in proteins [32,81,83,110].

ε-Amino groups in proteins.   Deprotonated surface ε-amino groups in proteins and polypeptides are readily alkylated by reagents VII, VIII, and XXVII. Another way of introducing a spin fragment into a lysine group is synthesis of a Schiff base with keto radicals of type R followed by hydrogenation in the presence of sodium borohydride [115,142]. Reagents that can be used to block the ε-amino group of lysine have been described [30,32,110]. Spin label VII proved to be the most convenient. Experiments with lysozyme, albumins, and bovine hemoglobin have shown that different protein groups react with the radical at different pH values, at least two pH regions being discernible. These are the pH 5–8 region, in which only deprotonated histidine groups are alkylated, and the pH 9–12 region, where lysine residues are also alkylated [92]. Blocking of the histidine-15 group in lysozyme makes it possible to modify only the lysine residues. The spin label is not added to hemoglobin at neutral pH, where the ε-NH_2 groups on lysine are shielded by protons. However, when the pH is increased to 10–10.5, addition of the spin label, apparently to the deprotonated ε-NH_2 lysine groups in hemoglobin, begins [92]. Human serum albumin at alkaline pH also takes on larger quantities of spin label VII than in neutral media.

Noncovalent interactions of spin probes.   Interesting new possibilities appear when noncovalent hydrophobic or electrostatic binding of spin-label compounds with biological structures is employed.

This applies primarily to hydrophobic derivatives of nitroxide radicals such as compounds XXXIII and XLIII (see p. 24).

The interaction between probes and the hydrophobic portions of the serum-albumin macromolecule was studied by various investigators [44, 115,143]. The albumin macromolecule tightly held the probes during gel chromatography on Sephadex. Probe XXXIII was selectively incorporated into the hydrophobic pocket of the active center of α-chymotrypsin [27]. Noncovalent binding also occurs on addition of spin-label analogs of substrates and inhibitors based on ATP, ADP, NAD, N-acetylglucosamine, and its derivatives, and so on [117–119,129] (see Chapter 6).

The electrostatic nature of the addition of paramagnetic copper-containing procion dyes of the 2RP type to proteins has been demonstrated [3,19].

X-Ray diffraction analysis of single crystals of lysozyme in the presence of the nitroxide analog of acetamide has made it possible to accurately locate areas of noncovalent binding of the reagent to the enzyme macromolecule [123].

The addition of spin labels to surface groups located outside the active centers of lysozyme [30,83,113,114,124], aspartate aminotransferase [33, 92], ferredoxin [114,145], myoglobin [32,83], hemoglobin [1,2,139], and nitrogenase [47] have only a slight effect on their biological activity and the physicochemical properties of the active centers (see Table 1). A list of several spin-labeled protein and enzyme preparations is given below. Modification of the active centers of enzymes by spin labels is described in Chapter 6.

Spin-labeled Group	Radical	Reference	Spin-labeled Group	Radical	Reference
Hemoglobin			Myosin		
β-93 SH	I	[1, 2, 92]	SH	I	[31, 147—150]
Same	II	[1, 3, 92]	Same	IV	[31, 148]
"	IV	[1, 3, 139]	"	VI	[31, 148]
"	VIII	[1, 3, 92]	Nitrogenase		
"	VI	[1, 3, 25, 82]	SH	I	[46, 47]
His, lys	VII	[92]	Same	IV	[46, 47, 94]
Same	XXVII	[92]	Fe component of nitrogenase		
Lys	XXVI	[92]	SH	I	[46, 47, 94]
Lysozyme			Same	IV	[46, 47]
His-15	VII	[30, 113, 114]	Fe—Mo component of nitrogenase		
"	XXVII	[113—114]	SH	I	[47]
"	IV	[83, 113, 114, 124]	Model iron-containing protein		
"	VIII	[83, 113, 114]	SH	I	[14, 94]
"	VI	[83, 113, 114]	α-Chymotrypsin		
His-15, lys	VII	[83, 113, 114]	Ser-195	XII	[65, 104]
Same	XXVII	[83, 113, 114]	"	XIII	[105]
Lys	VII	[83, 113, 114]	"	XIV	[105]
α-NH₂	VIII	[83, 113, 114]	"	XVI	[107, 108]
Asp-52	XXXI	[113, 114]	"	XVII	[106]
BSA			α-Chymotrypsin		
His	VII	[92]	His	XXV	[111]
His, lys	VIII	[92]	"	XXX	[111]
Same	VII	[94]	Ser-195	XVIII—XX	[109]
HSA			Ribonuclease		
SH	I	[44]	His-12	XXIII	[110]
Myoglobin			His	XXII	[110]
His	VII	[32, 83]	Lys	VIII	[110]
"	II	[32, 83]	Actin		
"	IV	[32, 83]	SH	XXXVIII	[121]
"	VIII	[32, 83]	Same	IX	[102]
"	VI	[32, 83]	Glyceraldehyde-3-phosphate dehydrogenase		
AAT			—	XXXVI	[103]
SH	I	[33]	Liver alcohol dehydrogenase		
Same	IV	[33]	—	XXXVI	[117—119]
"	VIII	[33]	SH	I	[120]
"	IV	[33]			

Continued

Spin-labeled Group	Radical	Reference	Spin-labeled Group	Radical	Reference
Lactate dehydrogenase			Pea ferredoxin		
SH	III	[96]	SH	IV	[144, 145]
Creatine Kinase			Corn ferredoxin		
SH	IV	[21]	SH	IV	[144, 145]
Phosphorylase b			Immunoglobulin		
SH	IV	[156]	His	VII	[146]
Cytochrome C					
—	XXX	[154]			
Lys	XXXII	[155]			

Note. BSA and HSA are bovine and human serum albumin; AAT is aspartate aminotransferase.

Table 1. Basic Properties of Lysozyme Preparations Modified by Spin Labels [114]

Lysozyme	Label	pH	Number of Labels per Macro-molecule	Assumed Location of Label	Relative Activity
0 (native)	—	7.0	—	—	1.0
1 (0 + iodoacetate)	—	6.2	—	His-15	1.0
1a	VII	6.2	0.95	"	0.94
1b	IV	6.2	0.85	"	0.6
1c	VI	6.2	1.1	"	0.96
2 (1 + VII)	VII	9.0	2.9	Lys-13, 96,98	0.9
3	VIII	7.0	1.5	His-15, lys-1	0.93
4	XXXI	3.9	0.55	Asp-52	0.45
5 (1 + IV)	IV	9.0	3.1	Lys-13, 96,97	0.92

The foregoing experimental material shows that different functional protein residues in a broad range of enzymes can be modified by spin labels of various structure. There is also the possibility of modifications that scarcely affect the basic biological functions and macromolecular structure of the systems under investigation.

Chapter Three
Double Paramagnetic Labels

In solving a number of important problems regarding enzyme action, it is necessary to know the distance between certain functional groups of the enzymes. For this purpose we must first decipher the three-dimensional arrangement of the groups in the active center. The study of allosteric transitions and other long-range effects also requires knowledge of the distances between the active center and the site of the allosteric effect. Experimental investigations of the structure of metalloenzymes, which contain several metal atoms per macromolecule, entail great difficulties.

A method employing double paramagnetic labels [3,14,19,20,22,81] was proposed and developed for deciphering the structures of biological objects, primarily enzymes. The method is based on the fact that different groups in a protein macromolecule can be modified by spin labels of identical or different structure. Under certain conditions the EPR spectrum of the spin-labeled proteins permits determination of the distance between the labels. Knowing the structure of the label, we can calculate the distance between the modified groups of the protein. This method has been further developed by other workers, both in the Soviet Union [23] and abroad [21,158].

In this chapter we present the basic principles of the method as well as results of its application to the study of the structure of myoglobin, lysozyme, hemoglobin, myosin, aspartate aminotransferase, albumin, nitrogenase and its components, antibiotics of the gramicidin type, and so forth. The data obtained by the double-label technique for the first three substances are compared with the results of X-ray diffraction investigations of these proteins.

THEORETICAL PRINCIPLES OF THE DOUBLE-LABEL TECHNIQUE

It is a well-known fact that under certain conditions the EPR spectra of paramagnetic centers show a fine reaction to the approach of other paramagnetic groups. There are two types of interactions between paramagnetic centers: (a) dipole–dipole interactions, which are due to the fact that the magnetic dipole of one paramagnetic group induces a local magnetic field at the site of another group and (b) exchange interactions, which are caused by the overlapping of the orbitals of unpaired electrons when paramagnetic particles are brought near to one another. In order to analyze the possibilities of determining the distance between two centers with EPR, we consider both types of interactions.

Dipole–dipole interaction. The magnetic field strength induced by a magnetic moment μ at a distance r is proportional to μ/r^3. Each spin in a system of interacting spins in a constant magnetic field will experience the additional effect of the local fields of the nearby spins. According to Eq. (1.1), this results in a spread of resonance frequencies and consequently in a change in the lineshape of the spectrum. There may be splitting, broadening, or a change in lineshape, depending on the specific situation.

The magnitude of the dipole–dipole splitting for a pair of isolated spins separated by a distance r is described by the equation [159]

$$\Delta H_{dd} = D_1(1-3\cos^2\theta), \qquad (3.1)$$

where $D_1 = (\frac{3}{2}g\beta/r^3)$ Oe, here β is the Bohr magneton and θ is the angle between the vector joining the spins and the direction of the magnetic field. The splitting of the EPR lines can be observed in magnetic single crystals in the dilute state in which there are definite and identical orientations for all of the pairs.

The interacting radicals in model and biological systems are generally oriented at random, and there is a whole set of distances between the dipoles. The g-factor anisotropy and the hyperfine structure anisotropy then have an effect on the EPR spectra, and there is often motion of the paramagnetic centers that averages the dipole–dipole interaction. Therefore, Eq. (3.1) cannot be applied directly to experimental spectra.

The analysis of EPR spectra is most conveniently carried out under conditions for which the molecular motion of the spins is either practically stopped (frozen solutions) or is slowed to the point at which it has no effect on the EPR spectra (when the centers are rigidly bound to very large macromolecules or in very viscous media).

The following methods, which are used to calculate distances between paramagnetic centers from the value of the dipole–dipole interaction, may be noted: (a) empirical calibration, (b) determination of second moments, (c) measurement of the dipole–dipole splitting, (d) analysis of the broadening of single spectral components, and (e) comparison of experimental and calculated spectra.

The method of empirical calibration is based on a comparison of the shape and parameters of the EPR spectrum of the substance under investigation and the spectrum of a specimen with a known arrangement of spins. The data on solid solutions of radicals and complexes or chemical compounds with a fixed structure, such as polyradicals, can be used as standard data.

The second moment (M_2) of a line is defined by the formula

$$M_2 = \frac{\int_0^\infty H^2 F(H)\, dH}{\int_0^\infty F(H)\, dH} - \left[\frac{\int_0^\infty HF(H)\, dH}{\int_0^\infty F(H)\, dH}\right]^2, \qquad (3.2)$$

where H is the magnetic field strength and $F(H)$ the integral absorption intensity.

The method involving second moments employs the van Vleck equation [160], which relates the dipole–dipole component of the second moment of a line (ΔM_2) to the parameters of the magnetic moments of several spins (N) and the distance (r_{ik}) between them

$$\Delta M_2 \doteq a\gamma^2 h^2 S(S+1)\frac{1}{N}\sum \frac{1}{r_{ik}^6}, \qquad (3.3)$$

where γ is the gyromagnetic ratio, S is the spin of the interacting particles, and a is a coefficient.

In the case of a polycrystalline sample of spins with $S = \frac{1}{2}$, the value of the average distance between the spins in a cubic lattice equals

$$r = 33.2/\Delta M_2^{1/6}\ \text{Å}, \qquad (3.4)$$

according to Eq. (3.3), if ΔM_2 is given in Oe^2. In the case of a radical pair, we can use the formula

$$r = 23.2/\Delta M_2^{1/6}\ \text{Å}. \qquad (3.5)$$

In practice the dipole–dipole component of the second moment of an EPR line for a system of interacting spins is calculated from the difference between the second moments for a given system (e.g., a polyradical) and for a system in which the interaction is known to be absent (a monoradical). The correctness of this approach has been substantiated [161].

In magnetic polycrystalline samples in the dilute state (biradicals in macromolecules, active centers of enzymes, radical pairs resulting from radiolysis, etc.) the dipole–dipole interaction may result in the splitting of the line by the value of D_1. In this case the distance between the pairs can be calculated from the formula

$$r = 30.3\, D_1^{-1/3}, \qquad (3.6)$$

which was obtained from Eq. (3.1) after substituting in numerical values [23]. If the spectrum has a component whose width varies as the spins are brought near to one another, the distance can be evaluated from the magnitude of the broadening ΔH_d (in Oe) according to the formula

$$r = 35.6\, \Delta H_d^{-1/3}. \qquad (3.7)$$

Double Paramagnetic Labels

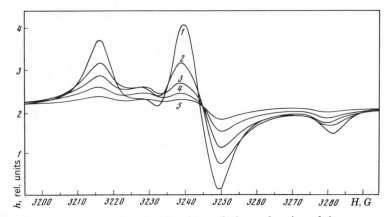

Figure 13. Theoretical spectra of a nitroxide radical as a function of the parameter D [21]: (*1*) $D = 0$; (*2*) $D = 3$; (*3*) $D = 10$; (*4*) $D = 30$; (*5*) $D = 100$. (See explanation in the text.)

This formula is rigorously valid only for cubic lattices [23,157], but the numerical coefficient is only slightly dependent on the three-dimensional arrangement of the spins.

In a number of cases EPR spectra can be calculated theoretically. The effect of the dipole–dipole interaction on the lineshape of nitroxide radicals has been analyzed [21]. The contribution of this interaction to the line-width was calculated from the formula

$$\delta H = D(1—3 \cos^2 \theta)^2, \tag{3.8}$$

where $D = (g\beta\mu^2/h \cdot r^6 \cdot \tau_1)$. Here τ_1 is the correlation time of the dipolar interaction, for example, the electron–spin relaxation time of the center causing the broadening. The theoretical curves obtained by Taylor et al. [21] (see Fig. 13) permit evaluation of r, if the other parameters in Eq. (3.8) are known.

A similar calculation was performed at the Institute of Organic Chemistry of the USSR Academy of Sciences by G. M. Zhidomirov and N. Ya. Shteinshneider. The spectra of nitroxide radicals were calculated with Gaussian lineshapes as functions of the magnitude of the dipolar interaction.

Exchange interaction. Bringing paramagnetic particles to distances that allow the overlapping of the orbitals of unpaired electrons results in an exchange interaction. This type of interaction is due to correlation and delocalization of the unpaired electrons, as a result of which each spin induces a finite spin density at the site of another spin. There are two types of exchange interaction: (a) dynamic and (b) static.

The physical meaning of the phenomena arising from exchange interactions may be visualized with the aid of a model that describes the behavior of spins in a constant magnetic field by means of a system of gyroscopes precessing about axes parallel to the field with identical resonance frequencies and phases.

Dynamic exchange interactions occur in solutions when paramagnetic particles encounter one another. In dilute solutions, in which the interactions occur only during short-lived encounters, the exchange effects manifest themselves as local fields whose fluctuations cause changes in the phases and frequencies of the precessing spins and thereby result in broadening of the EPR lines, the magnitude of the broadening being proportional to ν_c, the frequency of encounters between the paramagnetic particles. We thus have

$$\Delta H \sim \nu_c = k \cdot C, \tag{3.9}$$

where k is the rate constant of bimolecular encounters between the particles and C the concentration of radicals. Spin exchange is enhanced in concentrated (1–7 M) solutions of paramagnetic groups. This results in rapid relative reorientation of the spins in opposite directions due to the correlation between the unpaired electrons and in averaging of the electron–nuclear and electron–electron spin interactions, which are factors in the linewidth. In this case exchange narrowing of the spectral lines is observed. When there is an electron–nuclear interaction producing a hyperfine structure in one of the spin systems, the following sequence of changes will be observed in the EPR spectrum as the exchange is enhanced. First, the hyperfine components of the spectrum will broaden, and then, because of the averaging of the local fields during the exchange interaction, these components will be drawn together and fuse into one line. Further increase in the exchange frequency will result in exchange narrowing of this line. The direct spin exchange that occurs during interactions between the nitroxide fragments of bi- and polyradicals is the source of the specific nature of the EPR spectra of these compounds, which manifests itself as additional lines between the components of the triplet signal characteristic of the EPR spectrum of the monoradicals [162–165]. The intensity of these additional lines depends on the structure of the polyradical, the solvent, and the temperature. The theory of the EPR spectra of biradicals has been presented [162,163]. The theoretical treatment took into account the energy of the exchange interaction J. The quantity J characterizes the magnitude of the overlapping between the orbitals of the unpaired electrons in an encounter. The reason for the appearance of the additional lines is that in the overlapped state the spins of the unpaired electrons precess in the local fields of the nuclei whose electrons participate in the exchange.

At the limit $J/A_{iso} \gg 1$ (very intense exchange) the spectrum of the biradicals consists of five lines (a quintet), and the ratio between the intensities of the lines is $1:2:3:2:1$. This fits the delocalization of the electrons at both paramagnetic centers with a frequency $\nu \gg A_{iso}$. When J/A_{iso} lies between 0.3 and 20, the value of J can be determined from the position of the additional lines relative to the lines of the main triplet. In the case of $J/A_{iso} < 0.3$, the spectrum is practically indistinguishable from the triplet spectrum of the monoradical.

Exchange interactions occur in dilute solutions of biradicals when the nitroxide fragments approach each other as a result of intramolecular motion. The time modulation of A_{iso} with frequency ν_{ex} causes broadening of the additional lines by the quantity

$$\Delta\Delta H = \tfrac{1}{4}\sqrt{3}\gamma \cdot A_{iso}^2 \nu_{ex}^{-1}. \tag{3.10}$$

Measurement of the difference between the outermost and additional lines ($\Delta\Delta H$) permits determination of ν_{ex}, for example, from the formula

$$\nu_{ex} = 6.6 \times 10^8/\Delta\Delta H \; \text{sec}^{-1}, \tag{3.11}$$

which was obtained from Eq. (3.10).

Obviously, the EPR spectra can be transformed experimentally from a quintet to a triplet by varying the specific conditions, for instance, by slowing the exchange when the temperature is decreased. The EPR spectra of triradicals and tetraradicals show seven and nine lines, respectively, which correspond to delocalization of the electrons at three and four equivalent nitrogen nuclei.

Since the EPR spectra of biradicals are very sensitive to the state of the local molecular environment, spin labels based on them are very precise tools for investigating the topography of protein surfaces. Implanting labels on neighboring protein groups may give rise to a biradical or polyradical state. Examples of the application of biradical labels in the study of the active centers of enzymes are cited in Chapter 6.

The condition for a static exchange interaction between two spins separated by a fixed distance is the direct overlapping of the orbitals of the unpaired electrons. Such overlapping usually occurs at distances less than 7–8 Å. These distances may be increased when there are bridges between the spins, such as in complexes of paramagnetic metals with conjugated ligands.

The foregoing, of course, does not exhaust the entire range of spin-exchange interactions. We have considered the cases that have either been encountered in experiments with spin-labeled proteins or are necessary for comparison. Thus, for evaluating the distance between spin labels in proteins by EPR we may take advantage of dipole–dipole effects that cause broadening or splitting of EPR lines and an increase in the second moments of the lines, as well as spin–spin exchange interactions that cause both broadening and narrowing of the lines. By varying the experimental conditions we can choose among specific interaction mechanisms. For example, freezing solutions and increasing their viscosity should diminish the dynamic exchange effects and enhance the dipole–dipole effects. Additional information on the mechanism of interaction between spins can also be obtained by using labels with different lengths and flexibilities.

EMPIRICAL CALIBRATION OF EPR SPECTRA FOR DISTANCES BETWEEN SPINS WITH THE AID OF MODEL SYSTEMS

The following two systems may be employed for empirical calibrations: (a) liquid and rigid solutions of paramagnetic particles and (b) compounds with a known spatial arrangement of centers (e.g., nitroxide polyradicals).

Solutions of the paramagnetic copper-containing dye 2RP.   The procion dye 2RP has been used to investigate the topography of protein groups [19]. The structure of the dye includes a complex of the paramagnetic copper atom. Experiments designed to determine the dependence of the EPR spectra of radicals on their concentration have shown that the triplet broadening due to partial localization of the spin on the nitrogen atom is in approximate agreement with the theoretical calculations and the data obtained in experiments with other paramagnetic centers. The latter indicate that the following relationship should be fulfilled [23]:

$$\Delta H = 33.2 \ C_\mathrm{M} \ \mathrm{Oe}. \tag{3.12}$$

For example, at dye concentrations of 0.1 M the value of ΔH is 3–4 Oe, which should have some effect on the shape of the spectrum in the vicinity of the nitrogen triplet (Fig. 14).

Solutions of a nitroxide radical.   Solid water–glycerol solutions of nitroxide radicals have been extensively studied [23,166]. The experiments revealed good agreement between the empirical data and Eq. (3.12). The arbitrarily selected parameter d_1/d proved very sensitive to changes in concentration. The value of this parameter correlated well with other empirical and theoretical spectral parameters, namely ΔH°_0, the width of the central component, and δ, the width of an individual line [23]. A plot of d_1/d as a function of the average distance between the centers is shown in Fig. 15. It may be used as a calibration curve for analyzing data on spin labels in proteins. However, it is then necessary to introduce a correction into the value of r by multiplying it by 0.7, which takes into account the fact that

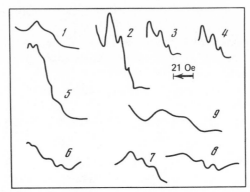

Figure 14. Fragments of the EPR spectra of the dye 2RP at 77°K in a water–glycerol matrix. Dye concentrations (in M): (1) 10^{-1}; (2) 5×10^{-2}; (3) 2.5×10^{-2}; (4) 1.25×10^{-2}; two molecules per albumin molecule at 20° in an aqueous solution: (5) 10^{-2}; (6) 10^{-1}; (7) 5×10^{-2}; (8) 1.25×10^{-2}; (9) five molecules per lysozyme molecule.

in solution each center is under the influence of a large number of neighboring centers, but in an isolated pair each center is under the influence of only one center. It should also be recalled that the value of d_1/d must be dependent on parameters describing the hyperfine structure anisotropy and the g-factor anisotropy, which vary somewhat as a function of the type of radical and the solvent. These variations are not important if the distances between the spins do not exceed 15–16 Å. When the distances are larger, the accuracy and reliability of this value depend on the extent to which the EPR spectra of the noninteracting radicals in the model system and in the system under investigation are identical.

Solid dilute solutions of polyradicals. The EPR spectra of biradicals 1–3 and 5–7, tetraradical 4, and monoradical R (Fig. 16) were recorded at 77°K in chloroform at concentrations of 10^{-5}–10^{-4} M for calibration of EPR spectra with respect to the distance between the spins and their relative position [22]. The EPR signals of the samples were recorded under conditions that eliminated saturation effects.

According to Fig. 16, as the nitroxide fragments are brought nearer to one another, the lineshapes of the spectra of biradicals 1–3 and the monoradical become increasingly different. At a distance of 16 Å the differences are markedly weakened and practically disappear when the centers are 18 Å apart. The spectra of biradicals 5–7 with n equal to 2 and 4 suggest a weak interaction between the unpaired electrons. At low values of n ($n = 2$ and 4) the nitroxide fragments are close. As n increases, the distance between the unpaired electrons increases, and at n equal to 6 and 10

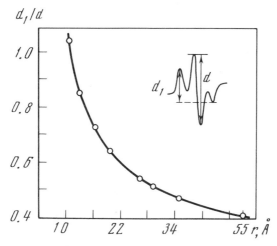

Figure 15. Dependence of an experimental parameter of the EPR spectra of the nitroxide radical $R(d_1/d)$ in rigid glassy water–glycerol solutions (77°K) on the mean distance (r) between interacting centers [23].

no interactions are evident in the spectra. This type of dependence of the average distances between unpaired electrons on n is also observed for dynamic exchange interactions [163–164].

Comparison of the spectra of biradicals *1–3*, tetraradical *4*, and polyradical *8* implies that the shape of the spectra depends not only on the average distance between spins, but also on their relative positions. Therefore, the values of the second moments of the spectra are conveniently used as an additional characteristic of the interaction (see Table 2).

Table 2. Second Moments (M_2) of the EPR Spectra of Polyradicals and Distances between Paramagnetic Centers

Radical (see Fig. 16)	M_2, Oe²	ΔM_2, Oe²	r_M, Å	r_{SB}, Å
1	405±10	68±14	11.5	12
2	345±10	8±14	$\lesssim 17$	16
3	340±10	~3	$\lesssim 18$	18
5 ($n = 2$)	340±10	~3	$\lesssim 18$	20
6 ($n = 4$)	345±10	~8	$\lesssim 17$	25
Polymer 8	337±10	200	10.8	10—12

Note. ΔM_2 is the contribution of the dipole–dipole interaction to the second moment; r_M is the distance calculated from Eq. (3.5); r_{SB} is the distance measured on Stuart–Briegleb models.

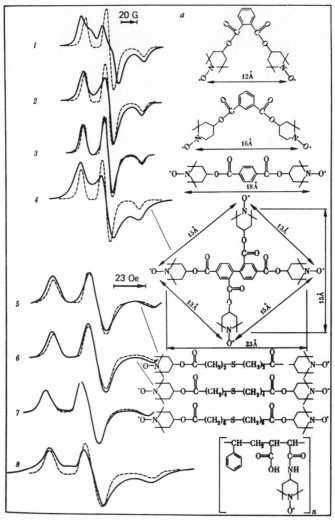

Figure 16. Chemical formulas of bi- (*1–3, 5–7*) and polyradicals (*4, 8*) and their respective EPR spectra [22,163]: (*a*) solvent, chloroform; radical concentration, 10^{-4} M; temperature, 77°K (rigid glassy matrices); solid lines, spectra of polyradicals; dashed lines, spectra of the monoradical R; (*b*) liquid solutions at 20°: (*1*) R_6–R_6; (*2–7*) $R_6OCO(CH_2)_nOCOR_6$, where $n = 0, 2, 4, 5, 6, 8$; *8–10*) biradicals *3, 2,* and *1* (see Fig. 16*a*), respectively.

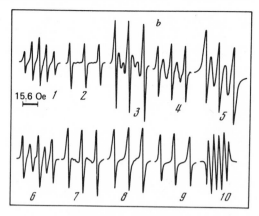

Figure 16. (*Continued*)

In Table 2 the value of ΔM_2 was calculated as the difference between the second moments of the biradicals and the second moment ($M_2 = 330 + 10$ Oe^2) of monoradical R. The mean square error in an analysis of several spectra was 10 Oe^2. If we assume that this quantity can be twice as large in the biological systems under investigation, the real possibilities of evaluating distances from second moments are limited to distances less than 16 Å. Models of biradicals show great conformational mobility, and r_{SB} was, therefore, measured under the assumption that the nitroxide fragments in biradicals *1–3* are radially directed, biradicals *5* and *6* are linear, and the polymer is twisted into a helix so that the nitroxide fragments are spaced with maximum freedom.

According to Table 2, the distances r_{SB} and r_M are similar for biradicals *1–3*, while for biradicals *5* and *6* there are noticeable differences due to their nonlinearity. Experiments with the biradicals have clearly shown that the dipole–dipole interaction in solid solutions of the radicals is observed at a distance less than 18 Å in the EPR spectra.

The distances between the unpaired electrons in the tetraradical were measured on a Stuart–Briegleb model, and the second moment was calculated from Eq. (3.3). It proved equal to 29, 32, and 53 Oe^2, respectively, for three different conformational states and to 54 Oe^2 for the distances cited in Fig. 16. A value of 53 ± 14 Oe^2 was obtained from the EPR spectra of the tetraradical.

The good fit between the results of the determination of r from EPR spectra and Stuart–Briegleb models permits use of the second-moment method for finding distances between paramagnetic labels added to protein macromolecules. Such data were also obtained by Kokorin et al. [23], who investigated the EPR spectra and a number of polyradicals.

Application of spin relaxation for evaluating distances between labels and paramagnetic ions. The evaluation of distances between labels and paramagnetic metal ions is difficult, because the spin relaxation of the paramagnetic ion, which averages the dipole–dipole interaction between the label and the ion to a considerable extent, is too great. In such a case analysis of EPR lineshapes makes it possible to determine only very small distances. It has been shown that the saturation curves of nitroxide radicals are more sensitive parameters than are the lineshapes [167–169].

Figure 17 illustrates the dependence of the amplitude of the low-field A component in the spectrum for the radical R—OH in alcohol at 77°K on the value of the microwave current in the detector (the saturation curve). The latter is conveniently characterized by the quantity I'_g (the microwave current in the detector at which half the maximum amplitude of the EPR signal is reached). The saturation curve is linearized in $(I_g/A)^2$ versus I'_g coordinates with the slope Δ. This makes it possible to find the parameter I'_g for curves that do not reach a maximum and to calculate the parameter Δ.

Figure 18 illustrates the dependence of the parameter I'_g for the radical R—OH in a 1:1 water–glycerol mixture at 77°K on the mean distance r between the molecules of the radical and the molecules interacting with them.

It is interesting to note that in the case of Mn^{2+} and $Fe(CN)_6^{3-}$ the EPR lineshape of the radical remains unchanged, while the parameter I'_g varies by almost an order of magnitude. In the case of Ni^{2+} the lineshape varies only for $r < 25$ Å. The saturation curves reflect the dependence on the effect of paramagnetic metals with greater sensitivity than do the lineshapes of the spectra. This can be explained in the following manner. In the case of a nitroxide radical at 77°K in the absence of paramagnetic ions, the rates of spin–lattice ($T°_1$) and spin–spin ($T°_2$) relaxation differ significantly,

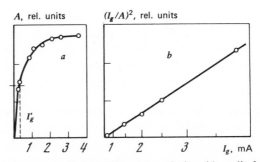

Figure 17. Saturation curve of the EPR spectra of nitroxide radical R in rigid glassy matrices (*a*) and the corresponding linearized plot (*b*) [170].

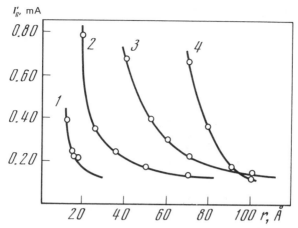

I'_g, mA

Figure 18. Dependence of the parameter of the saturation curves of the EPR spectra of nitroxide radical R (I'_g) on the average distance (r) from the radical to a paramagnetic metal [170]: (*1*) biradicals; (*2*) Ni^{2+}; (*3*) $Fe(CN)_6^{3-}$; (*4*) Mn^{2+}.

$1/T^\circ_1$ being much smaller than $1/T^\circ_2$. The times T_1 and T_2 for a 1:1 water–glycerol mixture at 77°K equal 3.8×10^{-4} sec and 10^{-7} sec, respectively (according to the results of Safronov and Muromtsev).

Since dipole–dipole interactions produce approximately equal increments [$1/T_1 = (1/T^\circ_1 + \Delta_1)$ and $1/T_2 = (1/T^\circ_2 + \Delta_2)$], T_1 is more sensitive to the effect of a paramagnetic ion than T_2. The course of saturation curves is determined by the values of T_1 and T_2, and they are more sensitively dependent on such effects than is the lineshape, which is regulated only by T_2. The different levels of the effects of paramagnetic metals are due to differences in their magnetic moments and relaxation times.

Analysis of the saturation curves makes it possible to expose different types of EPR signals of nitroxide radicals. Figure 19 displays the saturation curves for the following substances: (a) bovine hemoglobin modified with a spin label at the β-93 SH group and containing a free-label admixture, (b) the same substance following purification on Sephadex G-25, and (c) nitroxide radical R at a concentration of 5×10^{-4} M in alcohol. Curve *1* clearly indicates the existence of two types of radical added to the hemoglobin and coming under the influence of the heme and the residual free radicals.

Determination of the distance between the label and the paramagnetic center from the saturation curves requires a preliminary calibration with respect to I'_g or Δ. When there is an interaction between labels and heme, this calibration is performed on proteins with a known X-ray diffraction model, for example, myoglobin and hemoglobin labeled with labels of dif-

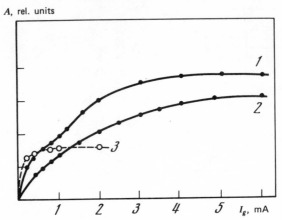

A, rel. units

Figure 19. Saturation curves of the EPR spectra of nitroxide radicals in heterogeneous systems (77°K) [170]. (See explanation in the text.)

ferent lengths [170]. The use of saturation curves for all of the proteins with paramagnetic centers that have been investigated with the aid of spin labels (nitrogenase, pea ferredoxin, mushroom tyrosinase, myoglobin, hemoglobin, and cytochrome C) has made it possible to safely establish that there is an interaction between the labels and the paramagnetic centers and under favorable conditions (when it is possible to evaluate the level of the effect of the paramagnetic metal on the radical in model experiments) to determine the distance between the radical and the paramagnetic metal.

Pulse EPR spectrometers [171] directly measure the spin–lattice relaxation time, but they generally have poor sensitivity, which can be increased by carrying out the experiments at low temperatures (10–20°K). Evidently, the further application of relaxation phenomena for investigations of the structure of proteins with labels will be based on the simultaneous, complementary use of pulse saturation and continuous saturation [171–174].

The solution of various structural and dynamic problems in biological systems will undoubtedly also be advanced by the ELDOR method [175–180]. The possibilities of ELDOR can be judged according to the following examples. The effect of the electron–nuclear interaction, which dominates at low radical concentrations and low temperatures, and spin exchange, which prevails at high concentrations and temperatures, on the parameters of the ELDOR spectrum has been established in solutions of a nitroxide radical in ethylbenzene [175]. The low-frequency intramolecular spin exchange in solutions of biradicals with $\nu_{ex} = 10^5 - 10^6$ sec^{-1}, which does not have a noticeable effect on EPR spectra, is, nevertheless, clearly detected by the ELDOR method [180].

Thus EPR offers a rich assortment of techniques for determining the distance between paramagnetic centers.

Determination of depth of paramagnetic centers relative to matrix surfaces. Two methods [43,167,168], the basic principles of which are represented schematically in Fig. 20, have been proposed for solving this problem. The first method involves comparing plots of the empirical parameters of saturation curves as functions of the distance r between radicals and relaxer ions randomly distributed in a rigid glassy matrix. In the case in which a saturated radical center is on the surface, the dependences of these parameters on r will differ only slightly from the corresponding calibration curves. If the center has penetrated the protein matrix to a depth of r_0, the corresponding curves must be shifted toward the ordinate axis because of the impossibility of bringing the relaxer ions closer than the distance r_0 from the paramagnetic center.

The second method is based on an analysis of the saturation curves of radicals randomly distributed over the surface of a protein globule containing the relaxer ion (e.g., the active center of an enzyme). In the case in which the relaxer center is located within a spherical globule, the saturation curves of the radicals must be characterized by one empirical parameter I'_g, which corresponds to the value of r_0. Displacement of the center gives rise to several distances and consequently to several parameters. This case requires special quantitative treatment, but it may be assumed that the limiting portions of the curves in $(I'_g/A)^2$ versus I_g^2 coordinates must correspond to the longest and shortest distances (r_0). A similar method had previously been suggested for finding the depth of paramagnetic centers relative to matrix surfaces from proton magnetic relaxation parameters (T_1) [181].

Empirical dependences of the saturation curve parameters I'_g and Δ of paramagnetic centers on matrix surfaces as functions of the concentration of an external paramagnetic atom at 77°K can be used to study the topography of protein surfaces [168].

It has been shown [167–169] that the value of Δ depends on the presence of paramagnetic metals. For example, in the case of Ni^{2+} in rigid water–glycerol solutions, we have the empirical relation

$$\Delta^{-1} = \Delta_0^{-1} + \beta[Ni^{2+}], \tag{3.13}$$

which is similar in structure to the relation between the relaxation rate and the concentration of paramagnetic metals (see p. 49). It is, therefore, natural to assume that the term $\beta[Ni^{2+}]$ is associated with the effect of the paramagnetic ions on the relaxation rate of the radical. Obviously β depends on the value of the solid angle Ω at which the paramagnetic ions can approach the radical. Therefore the quantity β is conveniently represented

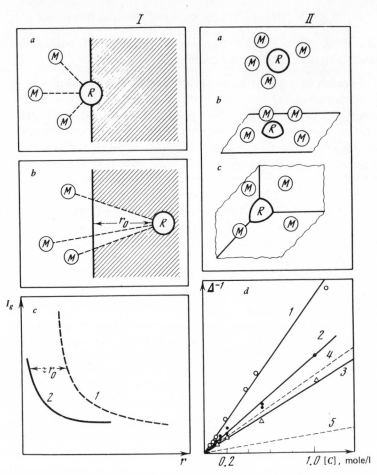

Figure 20. Techniques for determining the location of paramagnetic centers relative to the surface of a solid matrix [168,170]: (*I*) determination of the depth (r_0) of the paramagnetic center (R): (*a*) center on the surface; (*b*) center in the solid matrix; (*c*) dependence of the parameter I'_g on the average distance (*r*) to M (the relaxer ion) with a statistical distribution of M in the solution (*1*) and with the center localized in a solid matrix (*2*); (*II*) determination of the topography of the protein surface in the vicinity of R: (*a*) R is completely accessible, $\alpha = 1$; (*b*) R lies on a plane, $\alpha = \frac{1}{2}$; (*c*) R is at the intersection of three mutually perpendicular planes, $\alpha = \frac{1}{8}$; (*d*) dependence of the parameter Δ^{-1} (see text) on the concentration of M: (*1*) nitrosyl complex of nonheme iron (Fe^{2+} + xanthate + NO), $\alpha = 1$; (*2*) nitrosyl complex of the nonheme iron in nitrogenase, $\alpha = 0.67$; (*3*) nitrosyl complex of the nonheme iron in a model nonheme protein, $\alpha = 0.41$; (*4*) $\alpha = \frac{1}{2}$; (*5*) $\alpha = \frac{1}{8}$.

as the product $\beta = \alpha\beta_0$, where $\alpha = 1$ in the case of complete access to the radical in all directions and $\alpha = (\Omega/4\pi)$ in the case of steric hindrances.

By comparing the model and experimental systems we can determine α and thereby establish the level of access to the radical. This method has been tested on model spin-labeled proteins with known structures [168] and subsequently used to study the topography of a protein in the vicinity of the nitrosyl complexes of the nonheme iron-containing active centers of nitrogenase and a model nonheme iron-containing protein. According to the experimental values of α (Fig. 20), the nitrosyl complexes in the substances investigated are localized on the surface of the protein matrix.

This method gives correct results only in the absence of secondary phenomena (nonuniformity of the distribution of paramagnetic centers over the volume of the aqueous phase, selective addition to the matrix, chemical reactions, etc.). Therefore, an additional quantitative control is necessary in each case. It is thus wise to choose chemically inert complex ions with moderately hydrophobic ligands as relaxer ions and line broadeners and to add compounds that promote glass formation (glycerol, ethylene glycol, etc.) to the medium. Such precautions must be followed in all cases in which the double-label method is applied.

The types of interaction used to determine distances between the paramagnetic fragments of labels and the changes that correspond to them in the EPR spectra are shown schematically in Fig. 21. Methods employing double-spin labels, as seen heretofore, make it possible to determine distances between labels ranging from several Å up to 60–80 Å. This range practically covers the dimensions for individual enzyme molecules. As will be shown in Chapter 8, the use of electron-scattering labels makes it possible to expand to limits of the determinable distances to ≥ 100 Å.

We again stress that distances between the paramagnetic centers of the labels rather than between the functional groups themselves are determined in the variants of the method under discussion. Additional theoretical assumptions and further experiments are needed to determine the distances between the groups.

STUDY OF PROTEIN AND ENZYME TOPOGRAPHIES

The theoretical possibilities of the method and the encouraging results from model systems have not eliminated the need for independent tests of the applicability of the method for evaluating distances between protein groups. Lysozyme, myoglobin, and hemoglobin, whose structures in the crystalline state are known from X-ray diffraction data, have been used for this purpose. The method has since been used to evaluate the distances between functional groups in myosin, aspartate aminotransferase, serum albumin,

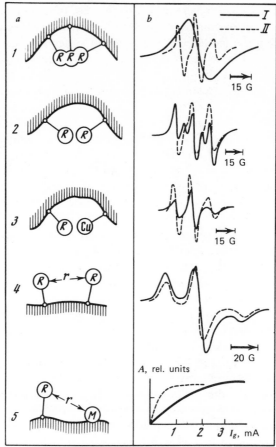

Figure 21. Evaluation of the distance between different portions of a protein matrix with the aid of double spin labels: (*a*) types of spin–spin interaction between labels; (*b*) corresponding changes in the spectra of the interacting labels; (*1*) exchange interaction between a group of nearby nitroxide radicals (295°K); (*2*) dynamic interaction between a pair of nitroxide radicals (315°K); (*3*) dynamic exchange interaction between nitroxide radicals and a complex of a paramagnetic metal, resulting in partial quenching of the signal of the radical (295°K); (*4*) static dipole–dipole interaction in frozen matrices (77°K); (*5*) variation in the shape of the saturation curves of nitroxide radicals as a function of the presence of paramagnetic ions (77°K); (*I*) interacting; (*II*) noninteracting labels.

nonheme proteins (nitrogenase and its components, ferredoxin), and so forth.

Lysozyme. Lysozyme is a compact protein with a comparatively low molecular weight and a known three-dimensional structure (Phillips model). This protein has a number of groups suitable for modification by spin labels.

When lysozyme is reacted with a solution of the dye 2RP at pH 6.3, one molecule of the protein combines with four paramagnetic molecules. Precisely as in the case of the optical parameters, the EPR spectral parameters of frozen water–glycerol solutions are identical for the dye on the protein and under model conditions if there is no more than one label for every protein molecule. For example, the following values for the linewidths (ΔH) and hyperfine structure (A) were observed in the case of the dye in the matrix: $\Delta H_{Cu_1} = 86$ Oe, $A_{Cu_1} = 140$ Oe, $A_{Cu_2} = A_{Cu_3} = 190$ Oe, $A_{N_1} = 15$ Oe, and $A_{N_2} = 12$ Oe. The values for the dye on the protein are: $A_{Cu_1} = 130$ Oe and $A_{Cu_2} = A_{Cu_3} = 190$ Oe. In aqueous solutions at $20°$ and pH 6.3 with the number of labels per protein molecule (n) equal to four, we have: $\Delta H_{Cu_1} = 85$ Oe, $A_{Cu_1} = 135$ Oe, and $A_{N_1} = 22$ Oe. These values approximate the values characteristic of solid matrices. The addition of the dye to lysozyme appears to involve two anionic sulfhydryl groups and is accompanied by significant restrictions on molecular motion. The EPR spectrum of a sample at $77°$K and a protein concentration of 2.4×10^{-3} M is similar to the spectra of the dye in a matrix at a concentration of 10^{-1} to 2×10^{-1} M, which corresponds to an average distance of 21–25 Å. This value is plausible, if we take into account the fact that the molecular cross-sectional dimensions of lysozyme do not exceed 20 Å.

In another series of experiments lysozyme was modified with a nitroxide label (IV), based on iodoacetamide. Depending on the pH this label can be added either to the histidine-15 group or to lysine groups. The EPR spectrum of lysozyme with a single label on the histidine-15 group coincides almost exactly with the spectrum of a dilute solution of the label in water. At alkaline pH values (pH > 9.5), alkylating agents primarily block the lysine groups. Additional information regarding the site of the labels has been obtained by the double-label method. A copper-containing paramagnetic label with an active cyanuric chloride residue was added near the spin-labeled histidine-15 residue but at a distance sufficient to quench the signal resulting from direct exchange interactions during encounters. According to the Phillips model, this region contains the lys-13, lys-96, and lys-97 surface groups [182,183]. Another argument in favor of the possibility of blocking the surface lysine groups nearest to the histidine group is provided by experiments designed to add the label at pH 11. According to certain data [138], the histidine and ϵ-NH_2 lysine groups of proteins are alkylated under these conditions. The EPR spectrum of lysozyme at $77°$K with one label almost coincides exactly with spectra of dilute solutions of the label in water. Addition of second, third, and fourth labels is accompanied by an increasing deviation from the spectrum of the single label and an increase in the second moment (Fig. 22). Comparison with the spectra of model biradicals and tetraradicals under analogous conditions reveals that, when there are two labels on lysozyme, the spectrum fits a

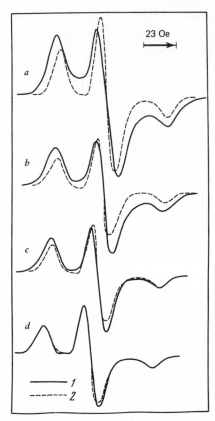

Figure 22. Electron paramagnetic resonance spectra of spin labels (*IV*) on egg lysozyme [113]: Here *n* is the number of labels per macromolecule and ΔM_2 the increment in the second moment of the spectra relative to the corresponding value for the monoradical (in Oe^2) with a protein concentration of 5×10^{-4} M: (*a*) $n = 4$, $\Delta M_2 = 50$; (*b*) $n = 3$, $\Delta M_2 = 35$; (*c*) $n = 2$, $\Delta M_2 = 10$; (*d*) $n = 1$, $\Delta M_2 = 0$; (*1*) label on the protein; (*2*) free label.

distance of 17–18 Å between the labels with paired spins [22]. The spectrum of lysozyme with four labels is markedly different from the spectrum of a tetraradical with spins separated by 13–15 Å, but approximates the spectrum of a biradical with spins separated by 15–16 Å. Following preliminary blocking of the his-15 group by iodoacetate, the spectra of lysozyme with two and three labels also fit the spectra of biradicals [83].

Stereoscopic photographs of the Phillips lysozyme model clearly reveal the possibility of a direct interaction between spin labels on the his-15 group and labels on the lys-13 and lys-97 residues as well as the approximately paired arrangement of the groups (lys-13 and his-15, lys-96 and lys-97, lys-97, and his-15).

The foregoing data imply that the treatment of lysozyme containing a blocked histidine group with label IV results in modification of the lys-13, lys-96, and lys-97 residues, which are located in the part of the protein globule farthest from the active center.

Incubation of lysozyme at pH 6.2 with label VIII results in 1,7 addition of the label to the lysozyme macromolecule. According to the EPR spectrum at 77°K, the labels are separated from each other by at least 18 Å [22]. Since at neutral pH maleimide reacts with histidine and lysine ε-NH$_2$ groups [138], we may assume that the objects modified in the present case are the his-15 group and the terminal α-NH$_2$ group on lys-1, which, according to the Phillips model, are at least 18–20 Å apart.

Introduction of spin-label XXXI into lysozyme in an amount equivalent to 0.5 labels for every macromolecule under conditions conducive to modification of the carboxyl group of aspartic acid [83,111,114] produces a proportional loss of enzyme activity. The EPR signal in this case fits the signal of the unreacted center. Comparison with certain reported results [184,185] suggests that the asp-52 carboxyl group, which is responsible for the enzymatic activity and is accessible to alkylation, is blocked.

Myoglobin. The data obtained with EPR [32,83] are in qualitative agreement with the X-ray diffraction model of myoglobin [186]. Analysis of the spectra of myoglobin with histidine groups modified by label V shows the almost complete absence of an interaction between the labels, which must be at least 18 Å apart. According to the Kendrew model, all four easily accessible histidine groups of myoglobin are separated from the heme group and from each other by at least 15 Å. The only exception is the pair in which the histidine residues are 8 Å apart. It appears that one of these groups remains practically unmodified under the conditions employed to obtain the spin-labeled protein.

Hemoglobin [99]. Titration of solutions of hemoglobin labeled at the β-93 SH group by radicals II and VIII with a paramagnetic dye that bonds electrostatically to proteins is accompanied by significant changes in the EPR spectra of the radical at dye concentrations ranging from 5×10^{-4} to 10^{-3} M, that is, at concentrations considerably lower than those needed to observe broadening of the spectra of the radical in model solutions $(2 \times 10^{-2} - 4 \times 10^{-2}$ M). As new portions of the dye are added, the intensity of the EPR signals decreases, the parameters ΔH and A_{iso} of the central part of the spectrum remaining practically unchanged. The effect is considerably weakened if the titration is carried out at pH 9. Since the radicals and the dye do not react chemically, the decrease in intensity may be associated with considerable broadening of some of the EPR signals.

The EPR spectra of spin-labeled hemoglobin with about seven molecules of dye per molecule of protein were recorded in a water–glycerol matrix at 77°K in order to ascertain the broadening mechanism. The EPR spectra of the dye-modified spin-labeled hemoglobin in the rigid matrix scarcely differs from the spectra of the free radical R and radical I added to hemoglobin under similar recording conditions. This fact apparently rules out the involvement of static dipole–dipole or exchange interactions. The empirical value of the linewidth in protein solutions of the dye is considerably greater than ΔH for the radical in solid matrices. This does not allow us to treat the intensity decrease and the broadening as results of the steric immobilization of radical I on the protein. The most likely mechanism is dynamic exchange broadening due to encounters between the paramagnetic ends of the nitroxide radicals and the copper-containing groups of the dye. This mechanism is possible if the labels are added to groups located at distances (r) that permit direct contact between the paramagnetic portions of the molecules. The dynamic nature of the effect is also confirmed by the fact that replacement of radical I by radical VI, which is more immobilized due to steric hindrances following addition to the β-93 SH group, significantly diminishes the effect of the addition of the dye 2RP to hemoglobin [20].

Thus the double-label method has demonstrated the presence of a positively charged ionogenic group located approximately 15 Å from the β-93 SH group.

Structure of active centers in nonheme iron–sulfur proteins: nitrogenase and its components, ferredoxin, and model proteins. The nonheme iron–sulfur proteins are a broad class of enzymes and carriers that catalyze such fundamental biochemical processes as oxidative phosphorylation, nitrogen fixation, hydroxylation, and photosynthesis [187–191]. All of these proteins have several atoms of iron in each macromolecule. Until recently there were no standard approaches to the solution of problems concerning the relative arrangement of the iron atoms [whether the atoms make up a single multinuclear complex (cluster) or whether they are distributed over the entire macromolecule].

One method proposed [3,14,45,49,93] for solving this problem includes chemical modification of cysteine residues that complex iron atoms in such proteins by a specific reagent (a mercurial) that contains a nitroxide radical. The spin label thereby dislodges the iron from the complex and replaces it. If the iron ions form a single multinuclear complex, the spin labels covalently bound to the cysteine residues must be close to one another.

In fact, the introduction of label I into the iron-containing centers of these proteins under optical and EPR control produces a singlet signal indicative of an intense exchange interaction between closely arranged spins (Figs. 23 and 24). The interaction disappears upon unfolding of the protein by urea. The effects observed are qualitatively identical for nitrogenase and its components, pea ferredoxin, and a model protein. However, proteins containing more than two iron atoms produce spectra suggestive of a more intense interaction following modification by spin labels. A similar

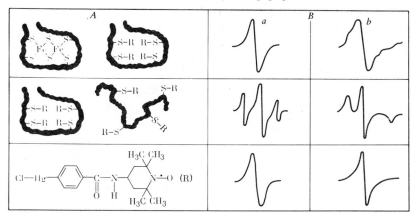

Figure 23. Method for investigating multinuclear iron-containing systems [3,93,94]: (*A*) schematic representations; (*B*) EPR spectra of a spin-labeled model iron-containing protein based on HSA in solutions at 195°K (*a*) and in rigid glassy water–glycerol matrices at 77°K (*b*).

approach employing a spin label based on an isocyanide has been used to study the arrangement of the surface iron atoms in nitrogenase [47]. Thus EPR has demonstrated the multinuclear structure of the active centers of nonheme iron–sulfur proteins. Figure 25 is a schematic representation of the structure of the iron-containing center of the model iron protein, which was obtained as a result of a whole series of investigations involving nuclear γ resonance, luminescent labels, EPR, and optical and chemical methods.

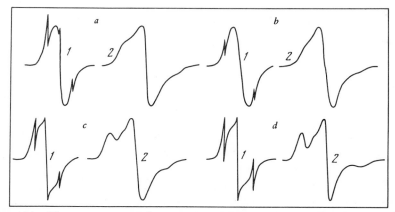

Figure 24. Electron paramagnetic resonance spectra of spin-labeled nonheme proteins (label I) [14,49]: (*a*) the model iron-containing protein; (*b*) Fe—Mo component of nitrogenase; (*c*) iron-containing component of nitrogenase; (*d*) pea ferredoxin; (*1*) at 295°K; (*2*) at 77°K.

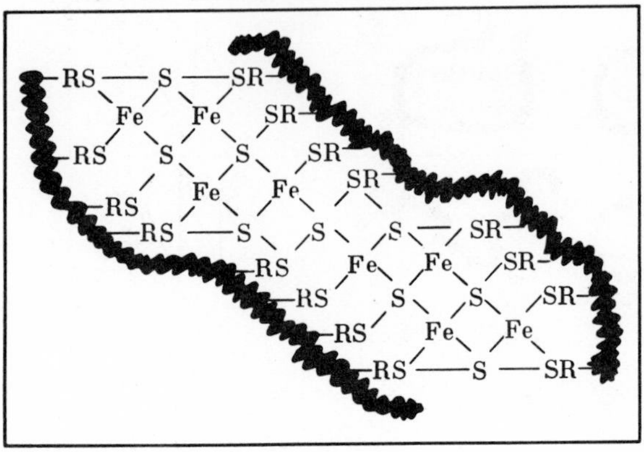

Figure 25. Structure of the nonheme multinuclear complex in the model iron-containing protein [94].

In another series of experiments carried out by A. V. Kulikov and L. A. Syrtsova at the Institute of Chemical Physics of the USSR Academy of Sciences, spin labeling was used to evaluate the distance between the sulf-hydryl group in the so-called "ATPase" portion of the active center and the nonheme iron. Addition of radical I to the SH group of the active center produces some alteration in the lineshape of the spectrum owing to an additional interaction, this alteration being most evident in the saturation parameter (Fig. 26). These changes fit the presence of a paramagnetic nonheme iron center about 15–17 Å from the nitroxide group. Considering the size of the spin label, we may assume that the SH group and the nonheme center are separated from each other by about 8 Å. This conclusion has also been confirmed with the aid of luminescent labels (see Chapter 8). It was found that the nonheme iron directly adjoins the ATPase center of nitrogenase.

Serum albumin. Titration of albumin preparations with spin-labeled lysine groups at various pH values with a solution of a dye whose molecules, as previously shown, are adsorbed on the cationic centers of the proteins produces three types of EPR signals from the nitroxide groups [20].

The first type is a broadened signal with splitting equal to 30 Oe due to nitroxide groups with strongly hindered rotation ($\nu \sim 10^-$ sec^{-1}).

The second type is a narrow signal with splitting equal to 16 Oe and a rotation frequency of 1.2×10^9 sec^{-1}. Such signals are broadened considerably upon addition of the first portions of the paramagnetic dye 2RP to the surface of the protein, and the number of signals capable of being broadened is practically independent of the pH in the pH 3–10 range.

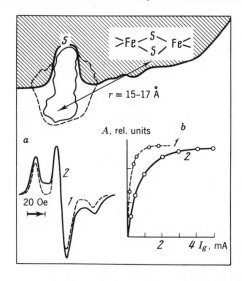

Figure 26. Structure of a fragment of the active center of nitrogenase in the vicinity of the SH group [47]: (*a*) EPR spectra of spin label I on the SH group of the enzyme; (*b*) saturation curve of the spectra; (*1*) model noninteracting systems; (*2*) spin-labeled nitrogenase.

The third type of signal is also assigned to slightly immobilized radicals. The difference between it and the second type is that the broadening upon addition of the dye is displayed at much greater dye concentrations (8 × 10^{-3} M as opposed to 2 × 10^{-3} M) and that the fraction of broadened signals varies significantly as a function of the extent of ionization of the groups with pK $\sim$ 6.5 and 10.

Thus double paramagnetic labels make it possible to distinguish between three kinds of lysine groups, which are apparently located in the vicinity of imidazole, histidine, and tryptophan groups, respectively, in human serum albumin.

The series of experiments with human serum albumin was carried out in such a manner that the test samples contained different numbers of 2RP molecules covalently bound to the protein molecules but maintained a constant total concentration of the dye following lyophilization [19]. The samples were prepared by dilution of preparations stained to different degrees with the pure protein followed by lyophilization. In this way conditions were created under which the interactions between the groups on different protein molecules were substantially weakened. By comparing the spectra of the dye on the protein with the spectra of the dye in a matrix (see Fig. 14), we could evaluate the effective label concentration on the surface and thus both the distance r between the labels on albumin at $n = 2$ and the mean distance $\bar{r}$ at $n > 2$. The value of r proved to be 28 Å, and the value of $\bar{r}$ varied as a function of n from 20 to 31 Å. The distance between paramagnetic centers undergoing a dipole–dipole interaction can also be determined from Eq. (3.7). The values of r from such a determination are equal to 18–22 Å, confirming the conclusion that the labels added to the proteins are separated by distances small enough for dipole–dipole interactions to develop.

These results allow us to approach the topography of a section of the albumin molecule. The distance between two dye groups that appear to be bound to the most accessible histidine residues of the protein is 28 ± 3 Å. These groups are situated 32.5 Å and 30 Å, respectively, from the tryptophan group [19]. This result is also given by the method of luminescence-quenching labels (Chapter 8).

The double-label technique has recently been used to evaluate the distance between the hydrophobic binding sites of human serum albumin and spin labels on the protein [115,143]. Hydrophobic spin probe XLIII competes with molecules of drugs and hormones (e.g., 6-methylprednisolone) for the hydrophobic portions of the protein and simultaneously somewhat alters the spectra of samples with spin-labeled histidine groups in magnetic frozen solutions in the dilute state. These alterations fit a distance of about 20 Å between the hydrophobic binding sites and the spin labels (spin label VII). The competition between the hormone and the hydrophobic probe is easily monitored by EPR, since the binding of the probe is accompanied by a sharp decrease in the EPR spectrum of the reaction mixture in solution. In frozen solutions the departing of the probe from the macromolecule into the solution is monitored according to the narrow intense singlet due to conglomeration of the radical molecules.

Other enzymes.   The macromolecule of rabbit myosin contains about 30 SH groups. Two of them are essential for ATPase activity: SH-I and SH-II. When the more reactive group, SH-I, is blocked by specific spin labels I and IV, the activity increases significantly [31], as in the case of alkylating agents. The EPR spectrum of a frozen solution of the spin-labeled protein differs only slightly from the spectrum of the monoradical. However, further addition of labels, producing a drop in enzymatic activity (addition to the SH-II groups), has a significant effect on the shape of the spectrum. This indicates that there is an interaction between labels separated by a distance $r < 17$ Å. Introduction of a paramagnetic Mn^{2+} ion into the active center of the enzyme causes only slight changes in the EPR spectrum of nitroxide label IV on the SH-I group, which correspond to $r > 14$ Å. The evaluations that have been carried out suggest that the activating and reaction portions of the active center of myosin are somewhat removed from one another.

Addition of a paramagnetic copper ion into the pyridoxal–phosphate active center of the enzyme aspartate aminotransferase has practically no effect on the EPR spectrum of label VIII added to the SH groups of the enzyme either in solution or upon freezing in liquid nitrogen [33]. This means that the pyridoxal–phosphate center and the label are separated from each other by a distance greater than 16–17 Å.

An approach that permits evaluation of the distance between the sulf-hydryl group in the active center and the Mn^{2+} ion in the Mn^{2+}-ATP–creatine kinase ternary complex has been proposed [21,157,158]. Analysis of the lineshape distorted by the dipolar interaction between the spin label on the SH group and the Mn^{2+} ion made it possible to establish that the distance between the centers is 7–8 Å.

Spin labels have been used to resolve the question concerning the relative arrangement of the heme groups of cytochrome P_{450} of the liver microsomal hydroxylating system [132]. According to EPR, electron spectroscopy, and kinetic measurements, the introduction of an isocyanide label into the heme group does not produce any significant changes in the spectrum of the system at 77°K. This indicates that the heme groups of cytochrome P_{450} are separated from each other by at least 17 Å.

Conformation of side-chain cyclic polypeptide groups. The first small polypeptide studied was the cyclic decapeptide gramicidin. The structure of gramicidin has been studied in detail [192,193]. A label based on cyanuric chloride (VII) was added to the side-chain ornithine NH_2, and doubly and singly labeled gramicidin derivatives were obtained [194]. The EPR spectrum of the monoradical derivative in solution indicates that there is rapid anisotropic rotation (Fig. 27). At − 196° the spectrum of this preparation coincides completely with the spectrum of the frozen monoradical.

In addition to the three principal lines, the spectra of solutions of the biradical derivative have two lines (Fig. 27) that indicate that there is an exchange interaction between unpaired electrons due to intramolecular encounters between the radical fragments. Comparison of the intensities of the additional lines leads to the conclusion that the frequency of encounters between the radicals (i.e., their approaching to within 6 Å) is considerably higher in ethanol than in chloroform.

The temperature dependence of the intensities of the additional lines has been used to evaluate the activation energy for the frequency of encounters in the biradicals. In the case of the ethanolic solution it was 5 kcal/mole, and in the case of chloroform it was 12–13 kcal/mole. The EPR spectra of the biradical derivative at − 196°, specifically, the lineshape and second moment [22], can be used to determine the distance between the nitroxide fragments. This distance proved to be 12.5 ± 2.5 Å. These results are in complete agreement with the pleated-sheet model of gramicidin, which was proposed in 1958 [195] and substantiated in detail in 1970 with the aid of a large number of spectroscopic and theoretical methods [192,195].

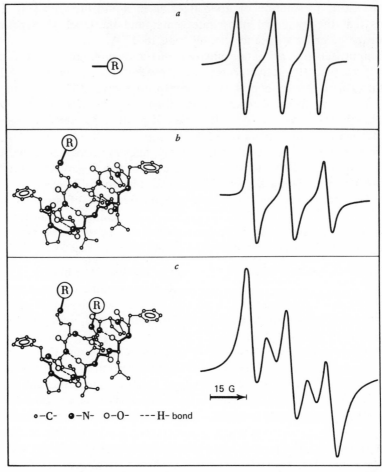

Figure 27. Determination of the conformation of the ornithine groups in gramicidin S with the aid of spin labels at 42° [194]: (*a*) R (radical VII); derivatives of gramicidin S: (*b*) monoradical; (*c*) biradical.

In fact, the characteristic features of the antibiotic structure are the rigid conformation of the polypeptide backbone, which is held in place by four intramolecular hydrogen bonds, and the arrangement of the side-chain groups of the ornithine residues on the same side of the plane of the ring, which facilitates their approaching each other in polar solvents. With respect to nonpolar solvents, additional hydrogen bonds between the NH_2 groups of the ornithine residues and the carboxyl groups of the ornithine acetyl side chains form in chloroform solutions [192]. Similar hydrogen bonds also form in the biradical derivative of gramicidin S, as indicated by the lower frequency of encounters between nitroxide fragments in chloroform than in ethanol. The difference between the activation energies (8 kcal/mole) corresponds approximately to hydrogen bonds of the $CO \cdots H$ type.

The examples cited show that spin labels can be successfully applied to the following: (a) study of the active centers of enzymes with known structures, (b) locating the sites of label addition relative to the active center of enzymes, (c) determination of the depth of paramagnetic centers relative to the surfaces of biological matrices, (d) the study of the structure and conformational state of various biological entities, and (e) the solution of other problems associated with determinations of distances between different parts of matrices.

Chapter Four
Investigation of Protein
Microstructure by the Spin
Probe–Spin Label Method

In the present chapter we analyze the new possibilities offered by the use of spin probes, that is, chemically inert paramagnetic species that diffuse freely in solution. The spin-probe technique has already been successfully applied to the study of spin relaxation, as well as the magnetic and other properties of paramagnetic complexes in solution [196–210]. Yu. B. Grebenshchikov and the author have suggested using this method to probe paramagnetic active centers of enzymes and to study the microstructure of proteins in the vicinity of experimentally added spin labels.

The value of the EPR linewidth ΔH_0 of a stable nitroxide radical in solution in a certain concentration range (10^{-3}–10^{-1} M) is related to the rate constant for exchange relaxation (k) upon encounters between the radical and a paramagnetic species by the simple equation

$$\Delta H_0 = 6.5 \times 10^{-8} \cdot k \cdot C, \tag{4.1}$$

where C is the concentration of the second paramagnetic species in M, ΔH_0 is expressed in Oe, and k is given in $M^{-1} \sec^{-1}$. The value of k for a given pair of paramagnetic species depends on their electronic structure, as well as the local viscosity and steric hindrances in the region of encounters. Therefore, following a preliminary empirical calibration, the method can be used in experimental studies of these factors in biological and other objects.

For example, Fig. 28 shows that questions regarding the location of paramagnetic active centers (transition-metal atoms and free-radical states) on the surface or within enzyme macromolecules are readily resolved with the aid of spin probes.

As we have seen in the preceding sections, the rotational diffusion parameters of spin labels covalently bound to protein groups reflect the state of the local topography of the protein matrix to some extent. Unfortunately, however, monitoring these parameters does not reveal the reasons for the slowing of the thermal motion of the nitroxide tips of the labels. In fact, such immobilization may be due to any of the following effects:

1. The nitroxide tip is embedded in the protein matrix.

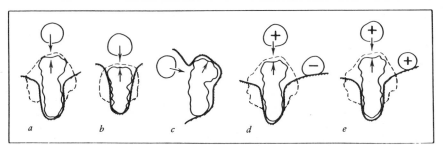

Figure 28. Mobility of spin labels and their accessibility to probes as functions of the topography and the presence of changes in the vicinity of labels [3]. The dashed lines show the effective volume for rotational diffusion; the arrows indicate the spin probes (circles) and labels; the shaded area denotes the surface of the matrix.

2. The spin label is almost entirely in viscous structured water.
3. The effective thermal motion of the tip of the label is slowed, because part of the "tail" of the label is held securely by the protein matrix.
4. The nitroxide fragment of the label is immobilized due to adsorption on the surface of the protein matrix.

Under certain conditions the application of spin probes makes it possible to determine which of these factors is actually operating in each specific case. Knowing the position of labels relative to the surface is exceptionally important in studying the local mobility of water–protein matrices with spin labels (Chapter 7).

Oppositely charged probes can be used to study the electrostatic charges near paramagnetic centers and then to "develop" the EPR spectra and analyze the heterogeneity of complex paramagnetic systems.

In this chapter we briefly discuss the theoretical basis of the method and consider our data and those of other workers on the effect of various factors (the electronic structure of paramagnetic complexes, the viscosity of the medium, and electrostatic effects) on the process of exchange relaxation during encounters between paramagnetic species. These data are used as model data in analyzing the behavior of paramagnetic particles in biological systems.

Spin relaxation in solutions of paramagnetic particles. The EPR spectra of solutions of radicals and paramagnetic complexes under specific conditions depend on their concentration and the concentration of the other paramagnetic particles. In principle, two types of spin interaction, namely, dipole–dipole and exchange interactions, can occur in solution. Some workers who studied the behavior of solutions of paramagnetic particles in detail [197–202,207] arrived at the conclusion that exchange interactions during encounters constitute the main factor in the EPR linewidth of nitroxide radicals in media of comparatively low viscosity. This conclusion is supported by the following results:

1. The relaxation rate falls with increasing viscosity of the solutions [197,207].
2. The spin–lattice relaxation times of the electronic spin show no appreciable effect on the linewidth in contrast to the theoretical predictions in the case of dipolar interactions [197].
3. The relaxation rate is greatly dependent on the ligand environment of the paramagnetic ion and to a lesser extent on the magnetic moment [197].
4. In many cases the rate constant for spin–spin relaxation is close to the number of encounters between the particles in solution.
5. The experimental data on the relaxation of paramagnetic ions with different structures are qualitatively in good agreement with the theories concerned with the delocalization of unpaired electrons over ligands and with the overlapping of electron orbitals during encounters between paramagnetic particles [197,210].

Additional information on the dominance of spin mechanisms has been obtained [42,211]. An analysis of the data on spin relaxation must include consideration of the translational diffusion of the paramagnetic particles, the rotational diffusion in a solvent cage [212], and the probability of intrinsic relaxation [197–210]. The probability of relaxation during an encounter between paramagnetic particles A and B is determined by the value of the exchange integral I_0. According to certain data [197,198], I_0 depends on the density of the unpaired electron at the sites of direct contact with the particles being encountered.

The value of I_0 is on the order of 10^{11} sec^{-1}, that is, close to the reciprocal of the lifetime of collisional complexes in solution (τ_1).

The probability of relaxation (P) during a bimolecular encounter is defined by the formula [38,197,201]

$$P = f \frac{I_0{}^2 \tau_1{}^2}{1 + I_0{}^2 \tau_1{}^2}. \tag{4.2}$$

where f is the nuclear–statistical factor and equals $\frac{2}{3}$ for nitroxide radicals. The value of P is also defined by the equation

$$k = z \cdot P, \tag{4.3}$$

where z is the number of encounters between the particles in solution and can be evaluated from conventional formulas for diffusion processes. According to this definition, the value of P includes a generally unknown factor that characterizes the steric hindrances.

A more general theory for exchange relaxation during encounters between paramagnetic particles in solution has been developed [204,206,210]. This theory takes into account the effect of such factors as the relaxation time of the electronic spins (T_1), the values of the magnetic spin numbers of the paramagnetic particles S_1 and S_2, and the value of the difference between the resonance frequencies of the spins (δ). The main conclusions from this theoretical treatment are listed in Table 3.

Table 3. Probabilities of Spin Relaxation of Radicals (P_1) and Complexes (P_2) during Encounters for Different Ratios between the Spin–Lattice Relaxation Time (T_1) and the Duration of the Encounters (τ_c) [210]

Relationship between I and δ	$\delta\tau_c$	$I\tau_c$	IT_1	$I^2 \cdot T_1 \cdot \tau_c$	P_1	P_2
				$\tau_c \ll T_1$		
	$\delta\tau_c \lesssim 1$	$I\tau_c \ll 1$	—	—	$\dfrac{I^2\tau_c{}^2\Sigma m^2}{2S_2+1}$	$^1/_2 I^2\tau_c{}^2$
For $I \ll \delta$	$\delta\tau_c \gg 1$	$I\tau_c \gg 1$	—	—	1 $(S_2$ even$)$	—
	—	—	—	—	$\dfrac{2S_2}{2S_2+1}$ $(S_2$ odd$)$	—
For $I \gg \delta$	$\delta\tau_c \gg 1$	$I\tau_c \gg 1$	—	—	1	1
	$\delta\tau_c \ll 1$	$I\tau_c \ll 1$	—	—	$^2/_3 S_2(S_2+1)I^2\tau_c{}^2$	$^1/_2 I^2\tau_c{}^2$
	—	$I\tau_c \gg 1$	—	—	$\dfrac{^2/_3 S_2(S_2+1)}{(S_2+^1/_2)^2}$	$\dfrac{1}{2\cdot(S_2\tau^1/_2)^2}$
				$\tau_c \gg T_1$		
—	—	—	$IT_1 \ll 1$	$I^2 T_1\tau_c \ll 1$	$^2/_3 I^2\tau_c T_1 S_2 \times$ $\times (S_2+1)$	—
—	—	—	—	$I^2 T_1\tau_c \gg 1$	1	—
—	—	—	$IT_1 \gg 1$	$I^2 T_1\tau_c \gg 1$	1	—

According to Table 3, the relaxation probability (P) during an encounter is determined to a great extent by the relationship between the time for spin relaxation, the lifetime of the complex, and the value of the exchange integral. Experimental criteria for strong $(P \sim 1)$ and weak $(P \ll 1)$ exchange have been found [205]. For example, in the case of strong exchange the temperature dependence of k must be determined by the temperature dependence of the number of encounters, that is, by the temperature dependence of the viscosity of the medium. In the case of weak exchange the value of k must be weakly dependent on the temperature because of the opposing dependences of z and P on the viscosity. In the case of strong exchange the value of k must be inversely proportional to the viscosity of the medium, and in the case of weak exchange, it must be independent of the viscosity.

Some of the conclusions of the theory were checked [205,210] in the case of interactions between nitroxide radicals and complexes of paramagnetic metals, and the validity of the theory was confirmed. This allows us to use the theory to formulate requirements for paramagnetic complexes in selecting a probe. Table 3 clearly shows that the best probe is a paramagnetic complex with very efficient relaxation, that is, with a value of P approaching unity and a small effective radius. In this case the experimental value of the relaxation rate constant will be proportional to the number of encounters. Equation (4.1) is rigorously valid for the homogeneously broadened EPR spectral line characteristic of comparatively low-viscosity solutions of paramagnetic particles. However, there is some basis to assume that this equation is valid to an approximation for inhomogeneously broadened spectra, which are superpositions of large numbers of individual lines. For example, analysis of certain data [23,166] leads to the conclusion that there is an approximately proportional relationship between the experimental width of the central component and the width of an individual line in EPR spectra of nitroxide radicals, even in the case of complete immobilization of their molecules. This means that Eq. (4.1) can be used to evaluate the relaxation-rate constant k for interactions between probes and spin labels on macromolecules, if the central component of the spectrum is used for the calculation.

Below there are experimental data on the effect of various factors on the spin-relaxation constant in solutions. These data will be used as a point of departure in interpreting the results of experiments with proteins.

Selection of spin probes.   The main criteria for selecting spin probes may be stated in the following manner. The probes must be readily soluble in water, chemically inert with respect to the functional groups of the biological systems and the nitroxide radicals, and not capable of being adsorbed on the surface of the matrices under investigation or of producing a concomitant EPR signal. Highly efficient exchange relaxation during encounters is an important quality for probes.

The following are the relaxation-rate constants ($k \cdot 10^{-8}$, in liter/mole·sec) for interactions between a nitroxide radical and several paramagnetic complexes in water and in 0.05 N H_2SO_4 [42,155]:

Compound	Water	Sulfuric Acid	Compound	Water	Sulfuric Acid
$VOSO_4$	8.8	—	$CoSO_4$	9.7	8.4
$CrCl_3$	8.8	—	$NiCl_2$	12·7	—
$Cr_2(SO_4)_3$	8.2	9.1	$NiSO_4$	13.4	12.7
$MnCl_2$	20.0	—	$Ni(NO_3)_2$	14.8	—
$Mn(CH_3COO)_2$	14.0	21.0	$CuCl_2$	15·0	—
$FeCl_3$	—	20.5	$CuSO_4$	14.7	15.0
$Fe(NO_3)_3$	21.0	24.5	$Na_2[Fe(CN)_5NO]$	0.0	—
$CoCl_2$	—	8.8	$K_3Fe(CN)_6$	12.2	—

The relaxation-rate constants ($k \cdot 10^{-8}$ 1/mole·sec) for interactions between nitroxide radicals and paramagnetic complexes in various solvents have the following values [42,155,199,204,205]:

Compound	Solvent	$k \cdot 10^{-8}$, (l/mole·sec)
$CrCl_3$	Py	17.5
$MnCl_2$	Py	35·7
$FeCl_3$	Py	26.3
$FeCl_2$	Py + H_2O	14.9
$CoCl_2$	Py + H_2O	8·2
$NiCl_2$	Py + H_2O	19.3
$CuCl_2$	Py + H_2O	30.6
Hemin	Py	24·0
Etioporphyrin F(III)	Py	25.7
$Mn(SCN)_2(\gamma\text{-}m\ Py)_4$	Py	38.5
$Fe(SCN)_2(\gamma\text{-}m\ Py)_4$	Py	6.1
$Co(SCN)_2(\gamma\text{-}m\ Py)_4$	Py	4.4
$Ni(SCN)_2(\gamma\text{-}m\ Py)_4$	Py	21.0
$Cu(SCN)_2(\gamma\text{-}m\ Py)_4$	Py	28·7
$Fe(AA)_3$	CCl_4	42.0
$Fe(C_5H_5)_2C_6H_2N_3O_7$	25% Acetic acid	15.0
Cu^{2+}	Water	14.0
$Cu^{2+}(H_2NCH_2CH_2NH_2)_2(H_2O)_2$	Water	19.0
$Cu^{2+}(H_2NCH_2NH_2)_2$	Ethanol	26.0
$Cu^{2+}[NH_2CH_2CH_2N(C_2H_5)_2]_2$	Water	10.0
$Cu^{2+}[(CH_3)_2NCH_2CH_2N(CH_3)_2]_2$	Water	35.0
$\{Cu^{2+}[N(CH_2\text{-}CH_2NH_2)_3]OH^-\}$	Water	25.0
$Cu^{2+}[(C_2H_5)_2NCH_2CH_2N(C_2H_5)_2]_2$	Ethanol	14.0

Continued

Compound	Solvent	$k \cdot 10^{-8}$, (1 /moll ·sec)
Cr^{3+}	Water	11.0
Mn^{2+}	Water	20.0
Fe^{2+}	Water	9.0
Co^{2+}	Water	9.0
Ni^{2+}	Water	11.0
$VO(AA)_2$	Cl	15.8
$Cr(AA)_2$	Cl	19.0
$Mn(AA)_3$	Cl	5.2
$Mn(AA)_2$	Cl	30.0
$Fe(AA)_3$	Cl	32.0
$Co(AA)_2$	Py	3.8
$Ni(AA)_2$	Cl	2.2
$Cu(AA)_2$	Cl	22.0
$Co(H_2O)_6(ClO_4)_2$	Water	9.0
$Co(CH_3OH)_6(ClO_4)_2$	CH_3OH	5.0
$Co(CH_3OCH_3)_6(ClO_4)_2$	CH_3COCH_3	2.7
$Co(CH_3COCH_3)_6(ClO_4)_2$	CH_3COCH_3	2.5
$Co(C_3H_7OH)_6(ClO_4)_2$	C_3H_7OH	3.0
$Co(C_3H_7OH)_2Cl_2$	C_3H_7OH	17.0
$Ni(AA)_2(C_6H_5H_2)_2$	$CHCl_3 + 3\%C_6H_5H_2$	13.0
$Ni(AA)_2(C_5H_5)_2$	$CHCl_3 + 3\%C_6H_5H_2$	23.0

Note. Py + H_2O, a 2:1 pyridine–water mixture, γ-m Py, γ-picoline; AA, acetylacetone; Py, pyridine; Cl, chloroform.

According to these data, the value of k is strongly dependent on the nature of the central atom and the ligand environment. The most suitable probes for the study of biological materials are potassium ferricyanide and dibenzenechromium iodide.

Potassium ferricyanide has a high relaxation probability ($P \sim 0.2$), forms a very stable ligand shell, is chemically inert, and has a short electronic relaxation time ($T_1 \ll \tau_c$). According to a certain experiment [211], the value of k is inversely proportional to the viscosity of water–glycerol mixtures, indicating that the value of P is independent of τ_c. This suggests that the value of k is proportional to the number of encounters in the case of ferricyanide–nitroxide radical pairs. Obviously, following the addition of a radical to a macromolecule, the inequality $T_1 \ll \tau_c$ can only be aggravated. This implies that the ratio between the relaxation-rate constants of a free radical in water (k_w) and on a protein (k_p) reflects the ratio between

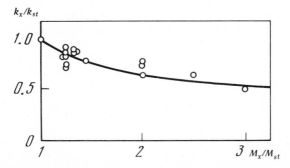

Figure 29. Dependence of the relative rate constant (k) for spin relaxation of nitroxide radicals with different structures on their relative mass (M) [211]. The indices st and x refer to the standard radical and the radical under investigation (see Tables 5–7).

the frequencies of encounters between the ferricyanide ion and the radical before and after it is bound to the macromolecule.

Thus potassium ferricyanide can be used to quantitatively study factors that influence the frequency of encounters, namely, the local viscosity, steric hindrances, the presence of electrostatic charges, and so on.

The dibenzenechromium ion has an even higher spin-relaxation probability during encounters with nitroxide radicals, its value approaching unity. This allows us to recommend this complex as a paramagnetic probe in low-viscosity media. However, unlike the preceding case, the relationship $T_1 \ll \tau_c$ might not be fulfilled in the case of viscous media.

Since the number of encounters is inversely proportional to the reduced radius (R) of the colliding particles, in order to compare the accessibility of a radical to paramagnetic probes in either an experimental or a standard system, we must introduce a correction that takes into account the difference in the value of R between the macromolecules and the radical and that has a limiting value of 2 (Fig. 29). It is, therefore, convenient to use the quantity $\mu_1 = (k_w/2k_p)$ (the local viscosity), which characterizes the increase in the viscosity and steric hindrances in the vicinity of an encounter between a probe and a radical on a protein, as the accessibility parameter.

Investigation of interactions between spin-labeled biopolymers and spin probes. Experimental values of $k_w/2k_p$ for proteins and other biochemical structures modified by spin labels of various lengths and flexibilities are listed in Table 4.

The following general trends can be detected by considering the experimental data obtained from studying globular proteins. In the case of radicals with relatively large distances from the polypeptide residue to the

Table 4. Relative Parameters for Rotational Diffusion (ν_w/ν_p) of Spin Labels on Proteins and Relative Values of the Relaxation Rate Constants ($k_w/2k_p$) [25,42, 81, 155]

Group	Label	Probe	pH	ν_w/ν_p	$k_w/2k_p$
—	R	Ferricyanide	7.0	1	1.0
—	R	Dibenzenechromium	7.0	1	1.0
Hemoglobin SH	VIII	Dibenzenechromium	7.0	730	1.7
Same	VIII	"	9.0	730	1.8
"	I	Ferricyanide	7.2	130	1.2
"	I	"	6.9	100	1.2
"	IV	"	8.2	100	1.8
"	VI	"	7.1	145	2.2
"	VI	"	5.0	220	1.0
"	VI	"	6.0	280	1.4
"	VI	"	7.0	380	2.2
"	VI	"	8.2	290	1.8
"	VI	"	9.0	250	1.5
"	VI	"	10.0	250	1.1
"	VI	"	11.0	100	0.8
Myoglobin his	VIII	Ferricyanide	7.0	55	0.9
"	VII	"	7.0	220	1.2
"	VIII	Dibenzenechromium	7.0	17	0.9
"	IV	"	7.0	55	1.0
"	VII	"	7.0	220	0.8
Myosin SH	IV	Ferryicyanide	7.0	1000	3.6
Same	IV	Dibenzenechromium	7.0	1000	4.3
"	I	Ferricyanide	7.0	50	0.7
"	I	Dibenzenechromium	7.0	50	0.8
Lysozyme his-15	VII	Ferricyanide	6.0	150	1.4
Albumin his, lys	VII	"	9.0	29	1.8
Silk fibroin his	VII	"	7.0	5000	20.0

Note. The indices "w" and "p" refer to radicals in water and on proteins, respectively; $\nu_w = 8.7 \times 10^{10}$ sec^{-1}, kw = 12×10^8 liter/mole·sec (for ferricyanide); $k_w = 28 \times 10^8$ liter/mole·sec (for dibenzenechromium).

piperidine ring, the value of $k_w/2k_p$ is close to unity. This means that the values of the local viscosity in the vicinity of the nitroxide group and of the viscosity of pure water are similar. In the case of shorter labels the value of $k_w/2k_p$ ranges from 1.5 to 2.5. Thus the nitroxide fragments of labels are fairly accessible to paramagnetic probes.

This result is in good agreement with the conclusions derived from an analysis of X-ray diffraction models of proteins. For example, the myoglobin model constructed by O. B. Ptitsin and his co-workers at the Institute of Proteins of the USSR Academy of Sciences according to Kendrew's

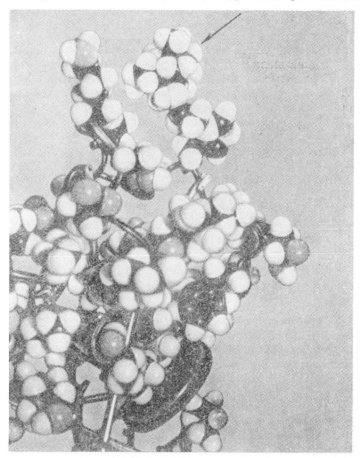

Figure 30.   Model of myoglobin (after Kendrew) with spin label IV on the histidine group. The black disc is the heme group, and the arrow indicates the label.

data clearly shows the possibility of the free approach of a probe to the nitroxide tips of labels bound to the surface histidine residues of the protein (Fig. 30). The Phillips X-ray diffraction model permits encounters between probes and labels added to the his-15 residue of lysozyme. A direct X-ray diffraction investigation of hemoglobin modified at the β-93 SH groups by label IV demonstrated that there is a conformation in which the nitroxide fragment is pointed out from the protein surface; this has been discovered independently with the aid of a probe.

While the rate constant for exchange relaxation k resulting from encounters with paramagnetic ions and the value of $k_w/2k_p$ for different radicals on proteins and in solution vary only insignificantly (by a factor of $\leqslant 2$–3), the rotational diffusion parameter ν varies at the limit by a factor of almost 5000 (see Table 4). Thus intramolecular rotations of labels on proteins experience additional steric hindrances, which have practically no effect on encounters with spin probes. This result is described qualitatively with the aid of the anisotropic rotation model for nitroxide-label fragments (Chapter 1). According to this model, there are interactions between the "leg" of the label and the water–protein matrix and weaker contact between the matrix and the fragment directed away from the protein surface.

Thus the application of a new method based on the simultaneous quantitative study of the rotational diffusion of paramagnetic labels on globular proteins and exchange relaxation during encounters with paramagnetic ions has led to the conclusion that in most of the globular proteins studied, the following conditions exist:

1. The nitroxide groups in the labels are directed away from the protein surface.
2. Rotational diffusion is slowed because of intramolecular steric interactions between the protein groups.
3. The local viscosity of the water in the vicinity of the nitroxide groups of the labels is close to the viscosity of water (in the case of long labels).

When the histidine groups in fibrillar macromolecules of silk fibroin [155] are modified in either the solid or dissolved states, the spin labels are inaccessible to ferricyanide, indicating the deeper implanting of the nitroxide tips in the polymer matrix. A similar picture has also been observed upon introduction of a spin label based on an isocyanide into the active center of cytochrome P_{450} in the microsomal hydroxylating system [132]. The active center of the cytochrome, which is buried in the protein–lipid matrix, is exposed only when it is unfolded. This process is clearly followed with the aid of the label-probe method.

Investigation of electrostatic charges in proteins with paramagnetic probes. The advances in the chemistry of stable nitroxide radicals [36] have made it possible to use these compounds with relative ease to study the effect of electrostatic factors on the process of spin relaxation. Such investigations permit quantitative testing of various electrostatic models of reactions in the liquid phase. Furthermore the results can be used for an empirical calibration of relaxation data as a function of the presence of electrostatic charges in the immediate vicinity of a paramagnetic center. Such empirical calibrations can then be used to detect anionic and cationic centers and to evaluate distances between functional groups in studies of the microstructure of proteins and other polymers with paramagnetic probes [81,155,211].

Investigations of the effect of electrostatic factors on spin relaxation in solutions employ stable nitroxide radicals of various structure and electrostatic charge (Table 5). Radicals not bearing a charge (Table 6) but similar in structure and molecular weight to the radicals listed in Table 5 are used as internal standards. This makes it possible to distinguish between the contributions to the value of the relaxation rate constant from electrostatic interactions and from changes in radical mass. Potassium ferricyanide, dibenzenechromium iodide, and iron(III) acetylacetonate have been employed as spin probes. The values of the rate constants for spin relaxation during encounters in solutions between nitroxide radicals differing in structure and charge polarity with the spin probes anionic ferricyanide and cationic dibenzenechromium are given in Tables 5–7. These tables also list the values of k for interactions between nitroxide radicals of one kind in solution.

The results of the experiments that have been carried out indicate the following qualitative laws. The values of the constants for a particular spin probe are greatly dependent on the structure of the substituent on the ring of the radical. The sign of the electrostatic charge on the substituent and its distance from the paramagnetic nitroxide group have a very significant effect on the value of the constants. A very strong electrostatic effect is observed as the sign of the substituent is varied in the case of the triply charged ferricyanide anion (Fig. 31). When the carboxyl group is replaced by an amino group, the ratio k^+/k^- varies by a factor greater than 15. Increasing the ionic strength produces a change in the value of k in qualitative agreement with Debye's theory [213].

Another factor having an appreciable effect on the value of the constant is the mass of the substituent. The experimental data in Tables 6 and 7 indicate that the values of k (for the same spin probe) are significantly dependent on the size of the substituent when neutral nitroxide radicals are employed. If a protein macromolecule serves as the substituent, the value of k decreases by a factor of two at the limit (see Fig. 29). The effect of the other properties of the substituent on the ring of the radical alters the value of k by no more than 20–25% [211]. The nature of the effect of the electrostatic charges and mass of the radical confirms our earlier conclusion that the relaxation rate is proportional to the number of encounters.

Theoretical calculations have been carried out for the following two model reactions between charged particles in solution: (a) a model of encounters based on the formula

$$k^{el}/k^0 = \frac{z_1 z_2 e^2}{kT D r} / e - 1$$

Table 5. Variation in the Rate Constant for Exchange Relaxation ($k \cdot 10^{-8}$, l/mole · sec) in aqueous Media as a Function of the Magnitude and Sign of the Charge on the Interacting Paramagnetic Ions [211]

Radical						
Formula	Number	Molecular Weight	pH	Ferricyanide	Dibenzenechromium	Same radical
O=⟨ ⟩N—O˙	I′	170	7	12.2	28.4	15.7
COO⁻ structure	II′	183	11	1.8	31.5	9.4
COO⁻ structure	III′	197	11	2.3	29.0	6.0
CH₂COO⁻ structure	IV′	211	11	2.8	27.3	6.0
O—CH₂CH₂COO⁻ structure	V′	243	11	3.5	22.7	5.3
CH₂CH₂—OOC / ⁻OOC structure	VI′	345	11	6.0	23.4	9.8

Table 5 (end)

| Radical | | | | | | |
Formula	Num-ber	Molecular Weight	pH	Ferricy-anide	Dibenzene-chromium	Same radical
(structure with NH$_3^+$)	VII'	156	4	16.8	20.0	12.6
(structure with NH$_3^+$)	VIII'	171	4	35.0	20.0	10.5

(the Debye formula [213]) and (b) a model that takes into account the electrostatic effects only in the collisional complex and is based on the formula

$$k^{el}/k^0 = \exp(-Z_1 Z_2 e^2/kTDr),$$

where k^{el} and k^0 are the rate constants for the charged and standard uncharged particles, respectively.

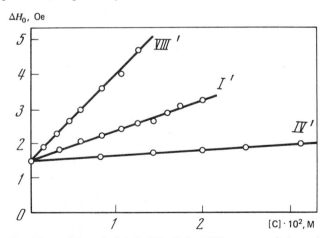

Figure 31. Experimental broadening (ΔH_0) of the EPR spectra of charged radicals in in solution as a function of the ferricyanide concentration [211]: I', IV', and $VIII'$-numbers of the radicals (see Table 5).

Table 6. Values of the Rate Constants for Exchange Relaxation ($k \cdot 10^{-8}$), l/mole · sec) in Aqueous Media for Radicals with Different Substituents in the γ Position [211]

Radical			Ferricy-anide	Dibenzene-chromium	Same Radical
Formula	Number	Molecular Weight			
	I′	170	12.2	28.4	15.7
	X′	224	10.8	24.8	13.6
	XI′	211	8.8	25.0	13.0
	XII′	213	9.1	24.5	13.0
	XIII′	340	7.7	21.2	12.2

Table 7. Exchange Relaxation Rate Constants for Radicals with Different Molecular Weights (the Spin Probe is Iron Acetylacetate, and the Solvent is Ethanol) [211]

Radical			$k \cdot 10^{-8}$, liter/mole.sec
Formula	Number	Molecular Weight	
$O={=}\langle\ \rangle N{-}O^{\cdot}$	I′	170	44.5
$\begin{matrix} HC{-}C{\nearrow}^{O} \\ \parallel \quad\ \ \searrow N{-}\langle\ \rangle N{-}O^{\cdot} \\ HC{-}C{\searrow}_{O} \end{matrix}$	XIV′	251	35.0
$C_{15}H_{31}{-}\underset{\parallel}{C}{-}O{-}CH_2{-}\langle\ \rangle N{-}O^{\cdot}$ (C=O below)	XV′	424	29.0
$Cl{-}Hg{-}\langle\ \rangle{-}\underset{\parallel}{C}{-}NH{-}\langle\ \rangle N{-}O^{\cdot}$ (C=O below)	XVI′	510	22.8

Two variants were calculated for each model. One was based on a dielectric constant equal to 80 for the water, and the other was based on a local dielectric constant that is dependent on the electrostatic field of the ion [211].

The value of r (the distance from the electrostatic charge on the radical to the paramagnetic group $> N{-}O^{\cdot}$) was determined for the radicals in Table 5 from Stuart–Briegleb models for different conformations. The diameter of the metal ions in the iron group equals 1.6 Å, and the diameter of the water molecules is 2.5 Å. The mean value of r for 26 cases (13 interacting pairs with two conformations in each pair) was equal to 9 Å.

Figure 32 presents curves described by the Debye equation with $D = 80$, the points being empirical data. An average calibration curve has been

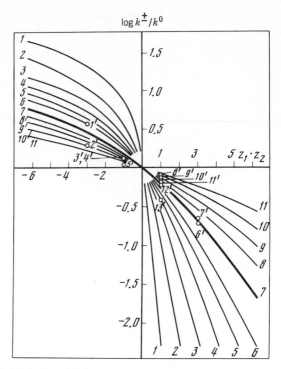

Figure 32.   Variation in the relative relaxation-rate constants ($k^{\pm}/k^0$) for nitroxide radicals in solution as a function of the product of the charges ($z_1 \cdot z_2$) of the colliding particles [211]: (*1–11*) theoretical curves calculated from the Debye equation for r equal, respectively, to 1, 2, 3, 4, 5, 6, 8, 10, 12, 15, and 20 Å; (*1'–13'*) experimental data (see Table 5).

constructed from these points. Within the range of the data, this curve coincides with the theoretical curve for $r = 8$ Å. At first glance this result seems surprising, since taking into account the local dielectric constant should yield a more correct value for k^{el}/k^0. If the electrostatic effect of the attractive and repulsive forces in the solvent cage are quantitatively taken into account, the value of k^{el}/k^0 should deviate even more from the experimental values. Apparently, the almost perfect fit between the experimental data and the values calculated theoretically from the Debye formula [213], which treats the electrostatic effect in a very simplified manner, is due not so much to its accuracy as to the fact that it effectively averages different opposing effects. The latter include the enhancement of the electrostatic effect due to decreases in the local dielectric constant, as well as enhancement and weakening of the electrostatic effects in the solvent cage as a result of the screening action of the electrolyte and the more uniform distribution of the electrostatic charge over the ligand molecules and the solvent. The effect of spin exchange at distances beyond the first coordination sphere due to delocalization of electrons is also possible.

Thus the empirical plots indicate that the Debye equation with $D = 80$ can be used to analyze experimental data and, in particular, to evaluate the approximate distance from the nitroxide group of a spin label on a protein to the nearest charged group, provided that this distance does not exceed 10–12 Å.

In the case in which radical VI is bound to the β-93 SH group of hemoglobin, the value of k_w/k_p is practically independent of the magnitude and sign of the charge on the paramagnetic probe (potassium ferricyanide and dibenzenechromium iodide) over a broad pH range [25]. This implies that the nearest charged groups on the protein in this case are separated from the nitroxide group of the radical by at least 12–14 Å. Similarly, in the case of myoglobin spin-labeled at the histidine groups, the values of k are relatively weakly dependent on the sign of the charge on the paramagnetic probe (see Table 4). A calibration curve can be used to evaluate the minimum distance to the nearest charged groups ($r \geq 12$ Å). In the case in which r is less than 12 Å, the constants k^+ and k^- must differ by at least threefold when the mean square error in the experiment is less than 15%.

Unlike the preceding cases, the data from titrations of spin labels on the "activating" sulfhydryl group in myosin with probes having different signs show a relative increase in the value of k_p^- over the value of $k_w/2k_p$. Here the value of $2k_p^-/k_w^0$ increases from 2.8 to 4.3 with increasing label length, while the value of $2k_p^+/k_w^0$ increases from 2.8 to 4.3 with increasing label length, and the value of $2k_p^+/k_w$ remains practically unchanged. These results can be attributed to the effect of a positive charge near the spin label.

Thus the use of spin probes with different charges appears to make it possible to detect charged protein groups in the vicinity of a bound spin label and in certain cases to evaluate the label-charge distance.

Investigation of the site of paramagnetic active centers in enzymes [215]. The dependence of experimentally observed relaxation parameters on the nature of complex-forming metals, as well as on the number and nature of the substituents in the porphyrin ring of complexes, has been studied. Porphyrin complexes of various metals appear in the active centers of a large number of proteins and enzymes, and the following substances were, therefore, selected for study:

Porphyrin Complex		$k \cdot 10^{-8}$ (l/mole·sec)	Final Concentration of Complex · 10^2, M
Mono-CET-TMP-Fe	(I'')	32.8	0.80
2,7-di-CET-TMP-Fe	(II'')	21.8	0.80
Tri-CET-TMP-Fe	(III'')	14.0	0.80
Tetra-CET-TMP-Fe	(IV'')	21.9	1.64
Mono-CET-TMP-Cu	(V'')	14.5	0.80
2,7-di-CET-TMP-Cu	(VI'')	9.8	0.80
Tri-CET-TMP-Cu	(VII'')	7.0	0.80
Hemin	(VIII'')	24.0	1.64
Etioporphyrin-Fe	(IX'')	25.7	1.64
Etioporphyrin-Cu	(X'')	27.1	0.44
Etioporphyrin-VO	(XI'')	15.3	1.64
Mesoformyletioporphyrin	(XII'')	14.3	0.80
Etioporphyrin-Ni	(XIII'')	0.0	0.80
Hemoglobin	(XIV'')	<1.0	0.60
Cytochrome C	(XV'')	<1.0	0.60

Note. The number of unpaired electrons in hemoglobin and Cytochrome C is five, in etioporphyrin-Ni it is zero, and in the remaining prophyrin complexes it is one. CET is the carbethoxy group, and TMP is tetramethyl porphyrin.

The interactions between radicals and paramagnetic prophyrin complexes are bimolecular processes and can be described by Eq. (4.1). The list of complexes includes values of the rate constants for the bimolecular exchange interactions involving porphyrin complexes of Fe(III), Cu(II), VO(II), and Ni(II), as well as hemoglobin and cytochrome C [215].

According to the data cited, in the case of the iron- and copper-containing complexes with conjugated ligands, including hemin, the values of the rate constants for exchange during encounters with nitroxide radicals are high, and such interactions effectively broaden the EPR lines of the radicals. However, under identical conditions there are no appreciable changes in the EPR spectra of solutions of a spin probe with methemoglobin and oxidized cytochrome C, which contain five unpaired electrons on each heme.

The results obtained imply that the active centers of heme enzymes in solution are within the protein globule. This is consistent with the Perutz model of crystalline hemoglobin [216].

The study of the interaction of radicals with hemoglobin and model metalloporphyrin complexes suggests that spin probes can be used to investigate the active centers of proteins and enzymes that contain paramagnetic metal atoms.

"Development" of the spectra of complex paramagnetic systems and study of the heterogeneity in the distribution of labels. Figure 33 illustrates the possibility of using the spin probe–spin label method to break down a complex EPR spectrum in which there is a superposition of signals from two radicals with different rotation frequencies and hyperfine linewidths (ΔH_0) between points of maximum slope. An immobilized radical is characterized by a fairly large linewidth, while a freely rotating radical produces a spectrum with narrow lines. Titration of a model consisting of a mixture of a freely diffusing nitroxide radical and spin-labeled hemoglobin with potassium ferricyanide makes it possible to easily isolate the signal of the broad component. The probe concentration sufficient to broaden and reduce the intensity of the narrow signal has little effect on the broad signal. A similar result has also been observed in titrations of spin-labeled silk fibroin samples, which produce a complicated signal from strongly and weakly immobilized labels, with ferricyanide (Fig. 33).

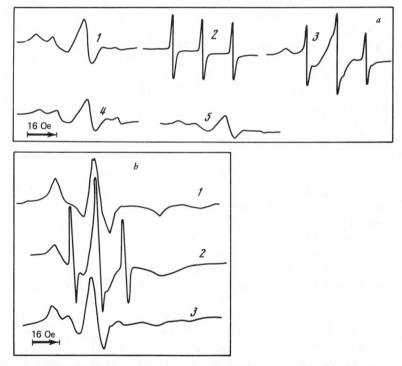

Figure 33. Quenching of narrow components in the EPR spectra of nitroxide radicals by the inert paramagnetic ferricyanide ion [42,81,155]: (a) control system: (1) radical VIII on hemoglobin (β-93 SH group); (2) radical VIII in solution; (3) mixture of solutions 1 and 2; (4) the same in the presence of 2×10^{-2} M ferricyanide; (5) the same in the presence of 3.5×10^{-2} M ferricyanide; (b) spin-labeled silk fibroin: (1) silk fibroin modified with label VII; (2) the same in the presence of excess radical; (3) the same as 2 but with an addition of 5×10^{-2} M ferricyanide.

The spin-probe method also permits analysis of the heterogeneity in the distribution of radicals over a macromolecule. The heterogeneity due to differences in the accessibility of the labels should appear as discontinuities on ΔH_0 versus $[C]$ plots. In fact, readily accessible spin labels have higher k values than do inaccessible radicals; therefore, the EPR spectra of such labels are broadened at lower probe concentrations.

Moreover in the case in which there is exchange between readily accessible and inaccessible label sites, the value of the constant for this exchange has an effect on the shape of the ΔH_0 versus $[C]$ plots. In particular, when the exchange is very rapid, the experimental values of k should become equal, and the discontinuity on the plots should disappear.

According to X-ray diffraction [139] and EPR [217] data, spin-label IV can occupy two positions relative to the hemoglobin macromolecule. In the first position the nitroxide tip of the label is rigidly embedded within the hydrophobic pocket. In the second position the tip of the label is pointed away from the protein globule. Obviously in the first position the nitroxide group is unavailable for direct encounters with a probe, while the second position permits direct spin exchange during encounters.

Figure 34 shows the dependence of ΔH_0 on $[C]$ for the iodoacetamide spin label on the β-SH group of human hemoglobin. This plot has a discontinuity at a probe concentration of 3.5×10^{-2} M, which qualitatively fits two different label positions. However, at higher probe concentrations at which the component h_1 for the readily accessible label vanishes completely because of the broadening, there is additional broadening of the central component.

The following are problems which, in principle, can be solved by the spin probe–spin label technique: (a) evaluation of the position of spin labels relative to a surface, (b) determination of the location of paramagnetic centers including active centers of enzymes, (c) detection of electrostatic charges in the vicinity of labels, (d) "development" of EPR spectra for complex systems, (e) study of heterogeneous systems, and (f) evaluation of the local viscosity in the vicinity of labels.

The spin-probe approach has several limitations. The exchange interaction between the probe (a paramagnetic complex in solution) and the spin label can be complicated by a chemical reaction, specific adsorption, dipole–dipole interactions, and so on. However, a thorough investigation that includes methods such as monitoring of the biological activity and the macromolecular structure of the object under study, analysis of the dependence of the linewidths on the reagent concentrations, employment of probes and labels with different structures, experiments carried out at different temperatures and viscosities, and comparison with the results from other methods permits isolation of the contribution of exchange interactions during encounters in each specific case.

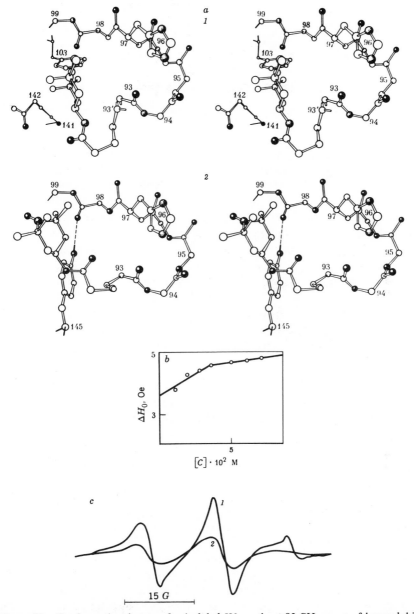

Figure 34. Conformational state of spin label IV on the β-93 SH group of hemoglobin [155]: (*a*) position of the nitroxide fragment of the spin label relative to the protein groups in the vicinity of the addition: (*1*) within the globule; (*2*) outside of the globule [139]; (*b*) broadening of the EPR spectra of the spin label as a function of the potassium ferricyanide concentration; (*c*) EPR spectra of spin-labeled hemoglobin: (*1*) without ferricyanide; (*2*) in the presence of 5×10^{-2} *M* ferricyanide.

Chapter Five
Nuclear Magnetic Resonance
of Spin-labeled Preparations

Nuclear magnetic resonance (NMR) has become one of the most effective tools for studying the structure and molecular motion of biological objects. The application of high-frequency and pulse NMR spectrometers, as well as techniques for specific chemical modification of proteins and other biochemical substances, especially selective isotopic substitution, in conjunction with biochemical and kinetic methods provides information on the fine structure and mobility of individual local segments of the systems under investigation, including active centers in enzymes, enzyme-substrate and enzyme-inhibitor complexes, membranes, nucleic acids, and so forth [218–227].

The present chapter presents the necessary elements of the theory of NMR for paramagnetic systems and the basic principles for studying proteins and enzymes by means of combined application of the methods of nuclear and electron magnetic resonance.

ELEMENTS OF THE THEORY UNDERLYING THE NMR
SPECTRA OF PARAMAGNETIC SYSTEMS

As we have seen in Chapter 1, the parameters of the EPR lineshape for nitroxide radicals and their dependence on the molecular motion are largely determined by dipolar and contact interactions between electron and nuclear spins. These interactions are effectively displayed in the NMR spectra of paramagnetic systems in the paramagnetic shift in the spectral lines and in the decrease in the spin–lattice (T_1) and spin–spin (T_2) relaxation times of nuclei possessing spin.

The magnitude of an electron–nuclear spin interaction is significantly influenced by a number of factors, particularly the distance between the spins (r), the values of the spin quantum numbers of the electrons (S) and the nuclei (I), the lifetime (τ_m) of the paramagnetic particles in the complex, the correlation time (τ_c^1) for the dipole–dipole electron–nuclear interaction, the density of the unpaired electron in the nucleus, and the intensity of chemical exchange between nuclei. Since direct contact interactions barely manifest themselves in biological systems, we shall limit ourselves to a discussion of phenomena associated only with dipole–dipole spin interactions.

The relaxation of a nuclear (e.g., a proton) spin ($H_1 \rightarrow H_2$) on complexation with a system containing a paramagnetic particle (e.g., the nitroxide radical R) may be represented by the following simplified scheme:

$$H_1 + R \xrightarrow{(1)} \underset{(M)}{H_1R} \xrightarrow{(2)} \underset{(M)}{H_2R} \xrightarrow{(3)} H_2 + R,$$

where the indices 1 and 2 refer, respectively, to the states before and after the act of relaxation. The overall process is composed of the purely kinetic steps of formation (1) and decomposition (3) of a complex and the magnetic relaxation of the protons (2). For the case most often encountered in biological systems, in which the rate is measured for the paramagnetic relaxation of protons in a substrate or inhibitor found in solution in excess relative to the number of paramagnetic active centers or spin-labeled molecules, the resultant rate constant (k^1) for the step-by-step steady-state relaxation of the protons can be determined from the expression

$$\frac{1}{k^1} = \frac{1}{[R]k_1} + \frac{1}{k_2} + \frac{1}{k_3}, \tag{5.1}$$

where $[R]$ is the concentration of the paramagnetic compound. The rate of the overall process can thus be determined either by kinetic $[R] \cdot k_1$, $k_3 \ll k_2$ or by relaxation $k_2 \ll [R] \cdot k_1$, k_3 steps. In certain cases additional experiments make it possible to ascertain the nature of the rate-limiting step. For example, the kinetic rate constants k_1 and k_3 should increase with increasing temperature according to an Arrhenius law, while the most likely result of the temperature effect on a dipole–dipole interaction in enzyme-substrate and enzyme-inhibitor complexes is a decrease in k_2. In experiments with enzyme-substrate and enzyme-inhibitor complexes, the formation and decomposition processes generally take place within less than 10^{-3}–10^{-4} sec under normal conditions, and the rates of the relaxation processes are determined by the rate of purely magnetic processes (the case of rapid exchange). Changes in the temperature and ambient conditions can alter the ratio between the constants in a given direction, and we can thereby experimentally determine the rates of the kinetic and relaxation steps individually.

In the case of rapid exchange the experimentally measured relaxation rates depend on the spin–lattice (T_1) and spin–spin (T_2) relaxation times in the paramagnetic complex M (T_{1M} and T_{2M}), on the ratio between the concentration of protons in the paramagnetic complex and their overall concentration (f_M), on the contribution of diamagnetic interactions and effects arising from direct encounters with paramagnetic particles (without formation of the complex M), and on the relationship between the resonance frequencies of the protons before and after complex formation. A complete mathematical expression that takes into account all of these factors has been analyzed by various investigators [124,157,228–230].

If the formation of the complex is not accompanied by a significant chemical shift, the value of T_{2M} is directly related to the structure and molecular motion of the paramagnetic complex and is calculated from the experimental data with the aid of the equation

$$\Delta \Delta H_{1/2} = \frac{f_M}{\pi} \frac{1}{T_{2M}}, \qquad (5.2)$$

where $\Delta \Delta H_{1/2}$ is the broadening of the proton-resonance line due to the formation of the paramagnetic complex. The quantity $\Delta H_{1/2}$ is defined as the width of a Lorentzian line at half its height and is expressed in hertz. The contribution of the paramagnetic complex to the increase in the longitudinal relaxation time (T_{1M}) is determined from the difference between the relaxation times in the presence and absence of the complex and consideration of the extent of complex formation.

According to the theory of Solomon and Blombergen [228–230], we have

$$\frac{1}{T_{2M}} = \frac{1}{15} \frac{S(S+1)\gamma_I^2 g^2 \beta^2}{r^6} \left[4\tau'_c + \frac{\tau'_c}{1 + (\omega_I - \omega_S)^2 \tau_c^2} \right.$$

$$\left. + \frac{3\tau'_c}{1 + \omega_I^2 \tau_c'^2} + \frac{6\tau'_c}{1 + \omega_S^2 \tau_c'^2} + \frac{6\tau'_c}{1 + (\omega_I + \omega_S)^2 \tau_c'^2} \right]; \quad (5.3)$$

$$\frac{1}{T_{1M}} = \frac{2}{15} \cdot \frac{S(S+1)\gamma_I^2 g^2 \beta^2}{r^6}$$

$$\left[\frac{\tau'_c}{1 + (\omega_I - \omega_S)^2 \tau_c'^2} + \frac{3\tau'_c}{1 + \omega_I^2 \tau_c'^2} + \frac{6\tau'_c}{1 + (\omega_I + \omega_S)^2 \tau_c'^2} \right], \quad (5.4)$$

where τ'_c is the correlation time for the dipole–dipole interaction, ω_I and ω_S are Larmor frequencies for the nuclear and electron spins, respectively, γ_I is the gyromagnetic ratio, g is the g factor, and β is the Bohr magneton of the electron. The quantities in the square brackets in Eqs. (5.3) and (5.4) are conveniently denoted by $f_2(\tau'_c)$ and $f_1(\tau'_c)$, respectively.

The function $\alpha \tau'_c / (1 + \omega^2 \tau'^2_c)$ passes through a sharp maximum; therefore, besides the first term in Eqs. (5.3) and (5.4), only those terms for which the value of τ''_c is close to the Larmor frequency ω are significant factors. In the case of a nitroxide radical bound to a macromolecule, the value of

τ'_c ranges from 10^{-7} to 10^{-8} sec. We can, therefore, simplify the expression for $f_2(\tau'_c)$ and write

$$f_2(\tau'_c) = \left[4\tau'_c + \frac{3\tau'_c}{1 + \omega_I^2 \tau_c^2} \right], \tag{5.5}$$

since the values $\omega_I = 6 \times 10^8 - 20 \times 10^8$ sec^{-1} and $\omega_S = 10^{10}$ sec^{-1} are substantially different.

The effectiveness of a dipole–dipole interaction in the general case can be determined by three parallel processes: (a) molecular rotation (with a correlation time τ_r), which averages the magnetic field induced by the electron spins at the site of the proton, (b) relaxation of an electron spin, which averages the interaction with an effective time τ_s, and (c) separation of the spins due to dissociation of the complex (with the time τ_m) (Fig. 35).

According to the equation for parallel processes

$$\frac{1}{\tau'_c} = \frac{1}{\tau_r} + \frac{1}{\tau_s} + \frac{1}{\tau_m}, \tag{5.6}$$

the effective correlation time is determined by the shortest time for one of the three processes cited above. Equations (5.3)–(5.6) reveal the unique possibilities for investigation of the fine structure of paramagnetic complexes by means of measurement of the rates T_1 and T_2. In fact, according to these equations, the values of T_{1M} and T_{2M} depend on three primary parameters, namely, the electron spin, the distance between the electrons and the nucleus under investigation (r), and the correlation time (τ'_c). If

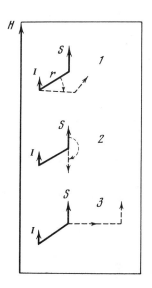

Figure 35. Types of effects resulting in disruption of the dipole–dipole interaction between the two spins **I** and **S**: (1–3) (see explanation in the text); r is the distance between the spins in the complex, and the dashed line indicates the direction of the motion of the spins.

the value of S is known, we need only make two independent measurements in order to determine these parameters. For example, we can measure T_1 and T_2 at one frequency of the NMR radiospectrometer or measure one of these values at different frequencies. In the case of substrates with known differences in r, it is possible to experimentally determine S for systems with unknown magnetic properties.

The most valuable data are those on the measurement of the distances r in enzyme-substrate complexes, since they permit construction of models without X-ray diffraction analysis; this is especially important for systems of high molecular weight. The prospects of determining τ'_c, whose values are related to molecular motion of the rotational diffusion type, are very enticing. In each case the method of calculation is determined by the specific problems and the initial data.

In the following section we shall consider specific examples of the application of NMR to the study of the structure of enzymes.

Investigation of the structure of enzyme-substrate and enzyme-inhibitor complexes with NMR. Beginning with the pioneer work of Jardetzky [218], attempts have repeatedly been made to use NMR to determine the structure of complexes of various compounds with enzymes from the magnitude of the diamagnetic effects. Measurement of the diamagnetic broadening of ligand molecules has made it possible to study the addition of antibiotics and sulfonamides to the serum-albumin molecule, the binding of the coenzyme NAD by dehydrogenases [218,219], and the addition of substrate analogs to chymotrypsin [222], aspartate transcarbamylase [224], lysozyme [222], and carbonic anhydrase [231]. For example, there has been study of the interaction of α- and β-N-acetylglucosamine with lysozyme, which manifests itself as an appreciable chemical shift for the inhibitor protons on contact with the aromatic tryptophan-108 residue [222].

Experiments with substrate analogs of pyruvate kinase revealed the unique possibilities of the application of Eqs. (5.1)–(5.6) to the study of internal rotation in various proton groups of the substrates when complexed with enzymes [225].

However, substantial progress in the application of NMR to the study of the structure of enzyme-substrate complexes was made only after enzymes modified with spin labels or paramagnetic metal ions, particularly manganese, were employed. The first advance in this direction was an investigation [128] in which the changes in the NMR lines for the histidine protons in the active center of ribonuclease were observed during the binding of phosphate label XLIV.

It is very often necessary to determine distances between the different ligand protons and paramagnetic centers in the enzyme. Since the value of the spin quantum number is generally known, the problem of calculating r reduces to a determination of which of three processes (relaxation of an electron spin, rotation of the macromolecule, and dissociation of the complex), which disrupt electron–nuclear dipolar interactions, is the most rapid [see Eq. 5.6)]. The values of τ_s range from 10^{-6} to 10^{-11} sec, those of τ_r from 3×10^{-11} to 10^{-6} sec, and those of τ_m from 10^{-5} to 10 sec (in the case of enzyme-substrate and enzyme-inhibitor complexes); therefore, the term containing τ_m is usually not used in calculations.

The rotational diffusion time of a spherical macromolecule with a molecular MW in an aqueous solution at room temperature can be evaluated from the formula

$$\tau_r = 7 \times 10^{-13} \cdot \text{MW sec}, \qquad (5.7)$$

which is derived under the assumption that Stokes–Einstein law applies with the aid of results from experiments with low-molecular-weight proteins [331].

In the case of nitroxide radicals, the value of τ_s can be evaluated on the basis of the following data. The experimental value of τ_s for aqueous solutions of Fermi's salt is 3.4×10^{-7} sec [124]. Experiments on the saturation of the EPR lines of a spin label on lysozyme yielded the inequality $\tau_s \geq 1.4 \times 10^{-7}$ sec [124]. Therefore the value $\tau_s = 2 \times 10^{-7}$ sec is reasonable for calculations for spin labels on proteins. Low-molecular complexes of manganese in aqueous solutions are characterized by τ'_c values ranging from 3×10^{-11} to 5×10^{-11} sec, which are due to the rapid rotational diffusion of the molecules (τ_r). The value of τ'_c for Mn^{2+} complexes of proteins is usually between 2×10^{-10} and 9×10^{-10} sec and is determined by the electron-relaxation time. One of the methods for experimental determination of τ'_c for complexes of Mn^{2+} is to use an equation that is a consequence of Eq. (5.4)

$$\varepsilon = \frac{f_1(\tau'_c)^* \cdot q^*}{f_1(\tau'_c) \cdot q}, \qquad (5.8)$$

where q and $f_1(\tau'_c)$ are, respectively, the coordination number and the correlation function for Mn^{2+} ions in water, q^* and $f_1(\tau'_c)^*$ are the corresponding values for a complex of unknown structure, and ε is the change in the rate of the spin–lattice relaxation of the complex with respect to the aqueous complex. Substitution of the numerical values given by Nowak and Mildvan [232] into the equation yields the quantity

$$f^*_1(\tau'_c) = \frac{\varepsilon \cdot 5.2 \times 10^{-10}}{q^*} \text{ sec.} \qquad (5.9)$$

It should be noted that the sixth power in the dependence of the quantities measured in the experiment on the distance r means that a twofold change in any factor in Eqs. (5.3) and (5.4) changes the value of r by 7%, and a tenfold change alters r by 40%.

The values of r can be calculated from the value of the paramagnetic broadening of the NMR line with the aid of the formula

$$r = 460 \left[S(S + 1) f_2(\tau'_c) \cdot \frac{f_m}{\Delta\Delta H_{1/2}} \right]^{1/6} \overset{\circ}{A}, \qquad (5.10)$$

which was derived from Eqs. (5.2) and (5.3). Here $\Delta\Delta H_{1/2}$ is expressed in hertz, and τ'_c is in sec. For example, according to certain data [124], in the case of a spin label on a comparatively rapidly rotating lysozyme molecule [$(\tau'_c) = 10^{-8}$ sec] in aqueous solutions at room temperature, we have

$$r = 23 \cdot [f_m/\Delta\Delta H_{1/2}]^{1/6} \overset{\circ}{A}. \qquad (5.11)$$

In the case of spin labels on macromolecules with molecular weights of $\geq 500{,}000$ or in very viscous media, the values of τ'_c and τ_s both equal 2×10^{-7} sec, and an analogous calculation yields the expression

$$r = 42[f_m/\Delta\Delta H_{1/2}]^{1/6} \overset{\circ}{A}. \qquad (5.12)$$

In the case of a macromolecule with a τ_r value close to the value of τ_s for the spin label, we have

$$r = 39(f_m/\Delta\Delta H_{1/2})^{1/6} \overset{\circ}{A}. \qquad (5.13)$$

Finally, iń the case of macromolecular complexes with Mn^{2+} the calculations can be performed with the expression

$$r = 23(f_m/\Delta\Delta H_{1/2})^{1/6} \overset{\circ}{A}, \qquad (5.14)$$

which was derived under the assumption that $\tau'_c = \tau_s \sim 4 \times 10^{-10}$ sec.

Alcohol dehydrogenase. The unique possibilities in the combined use of EPR and NMR for studying the structure of active centers of enzymes were first demonstrated in a thorough manner by Weiner and Mildvan [117–119] in the case of ternary complexes of liver alcohol dehydrogenase with a spin-labeled analog of the coenzyme NAD and ethanol. Spin label XXXVI, in which the nicotine residue has been replaced by a nitroxide fragment, competes with NAD and, according to the EPR spectrum, is tightly wrapped in the enzyme macromolecule. The entry of the label is accompanied bý a sharp increase in the relaxation time of the water protons that is dependent on the concentration of the specific substrate, ethanol. The NMR spectra of ethanol in the presence of label XXXVI are markedly broadened even when the concentration of the spin-labeled enzyme is 6×10^{-5} M (Fig. 36). In systems not containing the enzyme, similar broadening is observed only at much higher label concentrations (5×10^{-3} M). These results leave no doubt that this label is specifically bound by the macromolecule in such a manner that the nitroxide fragment occupies the site of the nicotinamide group in the native enzyme. Another very interesting result of their work was the unequal broadening for the various ethanol protons in the presence and absence of the enzyme. While in a solution of the label the NMR signals from the methylene group protons are broadened preferentially, in the presence of the enzyme the signals from the methyl protons are broadened most intensely. This implies that in the spin-labeled analog of the substrate complex the methyl groups are closer to the nitroxide fragments than are the methylene groups. Application of a computational method analogous to the one above makes it possible to determine the distance from the ethanol protons to the nitroxide fragment and to draw a schematic representation of the enzyme-substrate complex (Fig. 36). According to the scheme, direct transfer of a hydride ion is possible from the alcohol to the nicotine residue in NAD, as was assumed in earlier investigations. According to Weiner and Mildvan, the values found for the distances from the protons to the nitroxide fragment suggest the possible involvement of the tryptophan residue in this process. However, it appears to us that the values of these distances are somewhat overestimated (by about 20%) because of the choice of the underestimated value $\tau'_c \sim 2.8 \times 10^{-10}$ sec instead of $\tau_c = \tau_r < \tau_s \sim 2 \times 10^{-7}$ sec.

Analogous calculations have been performed for the specific inhibitors of the active center of alcohol dehydrogenase acetaldehyde and isobutyramide. Measurement of T_2 at different temperature makes it possible to determine the kinetic parameters for the exchange of ethanol in the ternary complex ($E_a = 5.4$ kcal and $\Delta S^{\neq} = -20.3$ eu.).

Phosphorylase and pyruvate kinase.   The complex problem of determining the relative position of the allosteric and active centers of phosphorylase b was solved by Bennick et al. [156] with the aid of a whole series of methods including measurement of: (a) the effect of Mn^{2+} in the active center on the NMR spectra of AMP (the activator) and glucose-1-phosphate (the substrate), (b) the fluorescence quenching of a chromophore label on the enzyme SH group, and (c) the interaction between an iodoacetamide label on the SH group and the ligand protons as well as Mn^{2+}.

Several variants of Eq. (5.3) were used to calculate r from the broadening of the NMR spectra of the substrate and activator (Fig. 37). The value $\tau'_c = \tau_r = 2 \times 10^{-8}$ sec was employed for nitroxide radicals on the protein, and the calculation for the manganese derivatives was performed with $\tau'_c = \tau_s$.

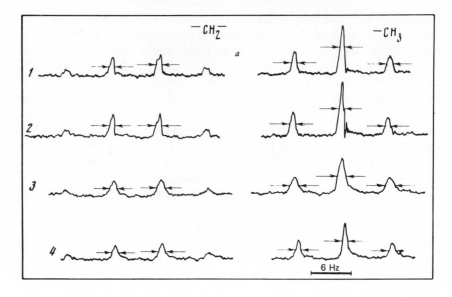

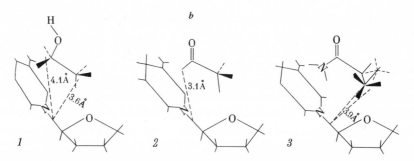

Figure 36. Results of an NMR investigation of spin-labeled alcohol dehydrogenase [119]: (*a*) NMR spectra of 5×10^{-2} M ethanol in water: (*1*) in the absence of paramagnetic labels; (*2*) in the presence of spin label XXXVI (1.65×10^{-4} M); (*3*) in the presence of spin label XXXVI (1.65×10^{-4} M) and the enzyme (6.1×10^{-5} M); (*4*) in the presence of spin label XXXVI (1.65×10^{-5} M), the enzyme (6.1×10^{-5} M), and NAD (2.3×10^{-2} M); (*b*) models of enzyme-substrate complexes of spin-labeled alcohol dehydrogenase: (*1*) ethanol; (*2*) acetaldehyde; (*3*) isobutyramide.

The results established the close arrangement of the allosteric and active centers and the location of the conformational changes under allosteric regulation. According to the diagram in Fig. 37, a molecule of the activator (AMP) is located between two paramagnetic centers approximately perpendicular to a line joining the nitroxide residue and Mn^{2+}.

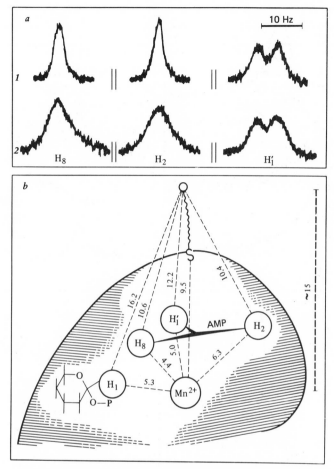

Figure 37. Results of an NMR investigation of spin-labeled phosphorylase b [156]: (*a*) NMR spectra of 2.19×10^{-2} M AMP in water: (*1*) in the presence of phosphorylase (4×10^{-4} M); (*2*) in the presence of spin-labeled phosphorylase (3.7×10^{-2} M); (*b*) model of a complex of the enzyme with a substrate analog and with the activator (AMP) (distance in Å).

Combined measurement of T_1 and T_2 for water protons and phospho-enolpyruvate analogs, as well as the broadening of the NMR spectral lines for the protons and phosphorus in phosphoglycolate in the presence of pyruvate kinase and Mn^{2+} enabled Nowak and Mildvan [232] to construct a detailed model of the enzyme-substrate complex. The NMR measurements clearly showed the changes in the substrate on inclusion of potassium ions in the active center.

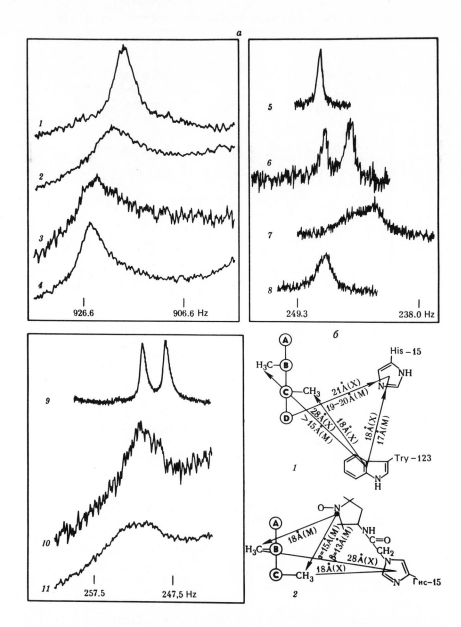

Figure 38. Results of an NMR investigation of spin-labeled lysozyme-inhibitor complexes [124]: (*a*) NMR spectra: (*1*) lysozyme (7 × 10⁻³ M), histidine-15 protons, pH 4.5; (*2*) lysozyme (7 × 10⁻³ M) + R$_{XL}$ (1.4 × 10⁻² M); (*3*) lysozyme (7 × 10⁻³ M) + NAG–R$_V$ (6 × 10⁻³ M); (*4*) lysozyme (7 × 10⁻³ M) + NAG–NAG–R$_V$ (6 × 10⁻³ M); (*5*) NAG (2 × 10⁻² M); (*6*) NAG (2 × 10⁻² M) + lysozyme (2 × 10⁻⁹ M); (*7*) NAG (2 × 10⁻²M) +

98

Lysozyme.   One NMR investigation of enzymes [124] employed a combined approach based on an analysis of the state of protein group protons during interactions between the enzyme and spin-labeled analogs of the substrate and a parallel analysis of the state of substrate protons in the case of a labeled enzyme. In the first series of experiments, performed by McConnell et al., egg lysozyme was selectively labeled at the his-15 group with spin-label IV on the side of the macromolecule furthest from the active center. The proton resonance lines of the specific substrates, N-acetyl-glucosamine (NAG) and its dimer (NAG–NAG) are appreciably broadened in the presence of the spin-labeled enzyme when compared to the proton resonance lines of the unlabeled diamagnetic enzyme. This made it possible for McConnell et al. to calculate the distances from selected protons on the α and β anomers of NAG to the nitroxide fragment by assuming that there is rapid exchange and that $\tau'_c = \tau_r = 10^{-8}$ sec. The experimentally observed differences in the distances from the methyl proteins of the anomers to the spin label on the protein were in good agreement with the differences that had been established with the aid of an X-ray diffraction model of lysozyme (Fig. 38).

The Phillips model permits two arrangements of the nitroxide fragment, outward from the protein surface and within the hydrophobic pocket. A scheme that allows exchange between the two arrangements fits the model best. In the other series of experiments, McConnell et al. used spin-labeled derivatives of acetamide, NAG, and NAG–NAG. The earlier X-ray diffraction analysis of lysozyme in the presence of the acetamide label XL showed that it can be bound to two subsites of the active center, most stably in the hydrophobic pocket in the vicinity of the tryptophan-123 group outside of the active center. The broadening of the resonance line for the N-2 proton on his-15 corresponds to a distance of 17 Å, which fits the distance of 18 Å measured directly from the model. It is interesting that the inclusion of NAG and NAG–NAG in the active center of the enzyme had no appreciable effect on the broadening. Although the accuracy of the NMR measurements in the present case was not very high due to the significant effect of the paramagnetic groups not bound to the enzyme, McConnell et al. assumed that no serious changes occurred in the distance between the tryptophan-123 and histidine-15 residues upon addi-

lysozyme–his-15-R_{VI} (2×10^{-3} M); (8) NAG (2×10^{-2} M) + chymotrypsin-ser-195-R_{XII} (2×10^{-3} M); (9) NAG–NAG (10^{-1} M); (10) NAG–NAG (10^{-1} M) + lysozyme (2×10^{-3} M); (11) NAG–NAG (10^{-1} M) + lysozyme-his-15-R_{VI} (2×10^{-3} M). The value of the chemical shift is given relative to tetramethylsulfoxide; (b) diagram of the distances in lysozyme-inhibitor complexes: (1) complexes of lysozyme with spin-labeled inhibitors; (2) complexes of spin-labeled lysozyme with inhibitors; (M) NMR; (X) X-ray diffraction analysis; (α and β) different anomers of NAG.

tion of NAG and NAG–NAG to subsites B and C of the lysozyme active center. Similar experiments with spin-labeled derivatives of NAG and NAG–NAG showed that the nitroxide fragment is located in the vicinity of subsite D. This conclusion was confirmed with the aid of the Phillips model.

Thus an investigation carried out on a protein with a precisely known structure has demonstrated that it is completely correct to use NMR to investigate the fine structure of spin-labeled biological entities.

Nitrogenase. The enzyme nitrogenase [50,191] efficiently reduces molecular nitrogen to ammonia under the influence of biological (ferredoxin, flavodoxin) and model (sodium dithionite) reducing agents only with occurrence of the coupled hydrolysis of a high-energy ATP bond, which activates electron transfer to the complexed nitrogen. The enzyme has two components: (a) component I, which contains 30–32 atoms of nonheme iron and two atoms of molybdenum, and (b) component II, which contains 2–4 atoms of nonheme iron.

Kinetic experiments with inhibitors and substrate analogs have led to the identification of three functionally different segments in the complicated active center of the enzyme: (a) the nitrogen-binding center I, (b) the reducing segment II, and (c) the ATPase or electron-activating center III.

Furthermore the ATPase center operates only in the presence of Mg^{2+}, Mn^{2+}, or other divalent metal ions and is blocked by reagents on the free sulfhydryl groups (there are two groups per macromolecule of molecular weight $\sim 350,000$), and the nitrogen-binding center is competitively inhibited by acrylonitrile, acetylene, and other compounds containing a triple bond and being less than 5–6 Å in length. Further physicochemical investigations (EPR, nuclear γ resonance, and optical spectroscopy) have shown that complexing ligands (CO and acetylene) that block the nitrogen-binding center do not affect the iron atoms and are most likely located on the molybdenum atoms of protein component I, while alterations in the structure of the iron-containing segments of both components under the effect of oxygen or the binding of iron by *ortho*-phenantroline inactivate the ATPase segment. Experiments involving the introduction of spin-label I into the sulfhydryl groups of the nitrogenase complex, followed by decomposition of the complex into its component parts, demonstrated that the nitroxide fragments occupy positions on both components. Spin-label and luminescent-label techniques, together with electron microscopy and nuclear γ resonance (at liquid helium temperatures), have established the cluster arrangement of the iron atoms in the Fe and Fe–Mo components [3,17,18,

46,50]. The formation of ternary complexes of nitrogenase with $ATP \cdot Mn^{2+}$ and $ADP \cdot Mn^{2+}$ was detected by the increase in the rate of longitudinal relaxation of the water protons $(1/T_1)$ [233]. Blocking of the free sulfhydryl groups on the enzyme eliminated the accelerated relaxation effect. In the native state at room temperature the enzyme produces no EPR signals. However, at 15°K there are signals with g-factor values of 2.015, 3.68, and 4.33, which correspond to high-spin iron [234,235]. The signals vanish when the enzyme functions under equilibrium conditions. Signals from Mo(V) and V(IV) are observed when the nitrogenase structure is altered by the presence of oxygen on acidification [50].

Probing the surfaces of oxidized and reduced nitrogenase preparations by measuring T_1 for the water protons has not revealed any paramagnetic centers that are accessible to solvent molecules [233]. Electron transfer from reduced component II to component I, monitored by EPR, occurs only in the presence of ATP [235].

All of these facts imply that the ATPase center of the enzyme includes the sulfhydryl group adjacent to the multinuclear iron-containing complex as well as adjoining portions of the multinuclear complexes of both components. The path of the electron from the external donor to the molecular nitrogen most likely lies along the Fe components, the multinuclear complex of the second component, and the molybdenum [50].

Nuclear magnetic resonance has been used to determine the relative positions of the ATPase and nitrogen-binding centers and to establish the magnetic state of the iron and molybdenum atoms in nitrogenase [135]. The specific substrate analog employed was acrylonitrile

$$\begin{array}{cc} H & H \\ \diagdown & \diagup \\ & C{=}C \\ \diagup & \diagdown \\ H & C{\equiv}N, \end{array}$$

which competitively inhibits the reduction of molecular nitrogen and is simultaneously reduced at its nitrile group in the nitrogen-binding center $(K_m \sim 10^{-2}$ M$)$. The experiment included measurement of the paramagnetic contribution to the broadening of the NMR lines of acrylonitrile as it enters the active center of the enzyme (E): (a) from the intrinsic paramagnetic centers of nitrogenase in the oxidized and reduced states $(E_{red}$ and $E_{ox})$, (b) from a spin label on the free sulfhydryl groups of the ATPase center $(E\text{-}S\text{-}R)$, (c) from a cluster of isonitrile spin labels that block the activity of the ATPase center by binding to the iron atoms $(E\text{-}Fe\text{-}R_n]$, and (d) from the ternary complex $E\text{-}ATP\text{-}Mn^{2+}$ (Fig. 39).

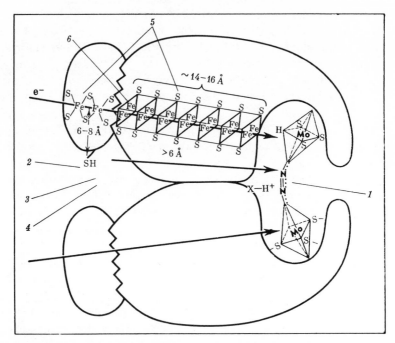

Figure 39. Model of the active center of nitrogenase [50]: (*1*) protons of acrylonitrile; (*2*) spin label I; (*3*) Mn²⁺-ATP complex; (*4*) Mn²⁺-ADP complex; (*5*) paramagnetic states of nonheme iron; (*6*) cluster of isonitrile spin labels; the arrows indicate the sites of addition of paramagnetic addition.   The diagram of the distances was constructed by analyzing a large body of data obtained with the aid of luminescent, spin, and electron-scattering labels, as well as with the aid of EPR, NMR, nuclear gamma resonance, and electron microscopy (see Chapters 3, 5, and 8).

Application of Eqs. (5.13) and (5.14) permits evaluation of the distance from the paramagnetic centers to the acrylonitrile protons in the active center of nitrogenase (see below):

Nitrogenase Complex	$\tau'_c \cdot 10^{10}$, sec	r, Å	Nitrogenase Complex	$\tau'_c \cdot 10^{10}$, sec	r, Å
E_{red}	~4	>7	E–ADP–Mn	~4	>6
E_{ox}	~4	>7	E–(Fe–R)$_n$	10^{2a}	>7
E–ATP–Mn	·~4	>6	E–S–R	10^3	>10

[a] Since the EPR spectra of the complex show a strong exchange interaction with a frequency of 10^8 sec⁻¹, the value of τ'_c should be close to the spin-exchange frequency.

According to the foregoing data, all of the paramagnetic centers in native and spin-labeled preparations of nitrogenase, including the multinuclear iron-containing complex, are separated from the acrylonitrile protons by distances greater than 6 Å. This seems to mean that the ATPase center (which includes the free SH group, the Mn^{2+}, and the multinuclear complex) and the nitrogen-binding segment are not in direct contact with each other. The iron atom in component I, which takes the electron from component II as a result of the activating effect of the hydrolysis of the pyrophosphate bond in ATP, is also outside of the nitrogen-binding center. The activated electron is transferred along the cluster in the substrate-binding center.

The examples cited above show that the NMR of paramagnetic systems opens up new prospects in the study of the fine structure and dynamic effects in proteins, enzymes, and other biological systems.

Chapter Six
Conformational Changes in Proteins and Enzymes

Even the first experiments with spin-labeled proteins [41] demonstrated the high sensitivity of the EPR spectra of spin labels to changes in the microstructure of proteins.

The present chapter presents experimental data on the application of spin labels in detecting various types of conformational changes, including: (a) comparatively large-scale denaturational changes under the effect of the pH, temperature, and various reagents, (b) allosteric effects, which are transmitted from one enzyme subunit to another, (c) generalized concerted transitions, which extend over protein globules when enzymes interact with substrates and inhibitors, and (d) changes in the state of various segments of the active centers of enzymes when specific reagents act on adjoining segments. Analysis of the conformational state and molecular mobility of the spin label itself as it interacts with a protein may be useful in studying the topography of the protein in the vicinity of the modified group.

Predenaturational phenomena in proteins under temperature changes can also be monitored with spin labels (Chapter 7).

Large-scale conformational changes. The thermal motion of spin labels on proteins is very sensitive to changes in the secondary and tertiary structure of proteins, if these changes affect the microstructure in the vicinity of the labels. The decomposition of hemoglobin into its subunits [101], the stretching of the polypeptide chains of hemoglobin [101], albumin [20], lysozyme [30], and aspartate aminotransferase [34], on ionization, thermal denaturation of hemoglobin [101], denaturation of proteins by urea [236] and dioxane [237], and other structural alterations, generally result in release of the labels and changes in the rotational diffusion parameters of the nitroxide fragments by factors of 100 or more [3]. The experimentally observed changes in the mobility of labels when the properties of the protein matrix are altered are largely dependent on the structure (length and flexibility) of their "legs." In those cases in which compounds with comparatively large distances from the polypeptide skeleton to the nitroxide group ($l \geq 12$ Å) are used, the molecular motion of the latter is only slightly dependent on the fine structure of the protein surface.

Spin label XXVIII, when placed on long and flexible lysine groups of polylysine and bovine serum albumin, has 20–50% of its mobility restored when the polypeptides are completely despiralized and form random coils.

The insensitivity of comparatively long spin labels on SH groups of bovine hemoglobin to decomposition of the protein into subunits and to despiralization at alkaline and acid pH has been observed [101]. Decreasing the length of the "leg" sharply increases the sensitivity of the molecular motion of labels, and thus the EPR spectra of the nitroxide group, to the topography of proteins near labeled protein residues. This phenomenon has been clearly revealed in the case of horse and bovine hemoglobin sulf-hydryl residues [101]. Acid and thermal denaturation of hemoglobin has a profound effect on the rotation of isomaleimide label VIII, whose ni-troxide end is tucked snugly in the hydrophobic pocket in native hemo-globin. Acid and base denaturation alter the parameters of the label by almost a thousandfold.

Ohnishi et al. [239] investigated the subunits of human hemoglobin modified at the SH groups with the fairly short spin label IV. Addition of this label to the single SH group of the α subunits, which according to the model in [216] is located outside of the contacts between the subunits in a fairly exposed portion, is accompanied by a decrease in immobilization ($\nu \sim 8 \times 10^8$ sec^{-1}). The EPR spectrum of the label is slightly dependent on the quaternary structure of the protein and changes somewhat when the hemoglobin is broken down into its subunits. According to the shape of the spectrum, this same spin label when added to the SH groups in the β subunits is held fairly rigidly between the subunits.

Release of the β subunits from the protein globule is accompanied by substantial restoration of the motion of label IV. The EPR spectrum of the labeled β subunits corresponds to a comparatively exposed portion of protein surface [216].

The effect of the label structure on EPR spectra has been observed by many workers, an overall trend being quite clear. An increase in the length of the labels on the same groups is accompanied by increasing restoration of their motion and decreasing sensitivity to various effects [2,3].

The sedimentation constants of human hemoglobin and its subunits labeled with radical I have been measured [239]. The addition of labels to the native hemoglobin molecule slightly increases the sedimentation con-stant. Prior addition of a label or PCMB (p-chloromercuribenzoate) to the α subunits does not prevent the formation of the quaternary structure and only slightly alters the sedimentation constant of the assembled molecule. However, labeled β subunits are held less tightly, and the sedimentation constant decreases by 0.4 units. Regeneration of the molecule is especially poor if both the α and β subunits are modified. The addition of label IV has a considerably milder effect than the addition of PCMB. The serious difficulty in assembling the molecule resulting from the modification of the sulfhydryl groups in the subunits is attributed to the fact that in the assembled molecule the SH groups are in the area of contact between the subunits. Spin labels on the sulfhydryl groups of the active center of lactate dehydrogenase are very sensitive to the effect of urea [96]. The mobility of the labels is restored at much lower concentrations than those required to alter the extent of spiralization as monitored by optical rotatory dispersion. Experiments on the denaturation of spin-labeled proteins have shown that the EPR spectral parameters of spin labels are highly sensitive to various effects and were used in preliminary studies before the attack of the method on the principal problems of molecular biology.

ALLOSTERIC EFFECTS IN PROTEINS AND ENZYMES

The first really significant result of the application of spin labels was the discovery of allosteric conformational changes in hemoglobin in solution. Spin labels made it possible not only to determine that the change occurs in the vicinity of the β 93 SH group at a distance of about 15 Å from the site of the reagent effect (the heme group), but also to reveal the details of this interesting phenomenon.

Allosteric effects were first clearly observed by McConnell et al. [28,238] in the case of spin-labeled horse hemoglobin. At pH values close to neutrality the EPR spectrum of this label fits fairly strongly hindered rotation. When oxygen is bound in the active center of hemoglobin, the spectra undergo definite changes (Fig. 40). The lateral components of the spectrum are markedly broadened, and splitting appears. All this undoubtedly indicates additional slowing of the rotation of the nitroxide group. It is clear that portions of the protein are contracted as a result of the allosteric effect, which is transmitted from the heme group to the vicinity of the label. Experiments with the individual subunits have shown that the additions of O_2 causes the hemoglobin β subunits to pass into a more compact conformation. Careful measurements of the dependence of the EPR spectra of labels in hemoglobin on the O_2 pressure revealed a slight deviation from the isobectic rule, which, in the opinion of McConnell et al. [238], is due to a change in the binding constants of the second, third, and fourth O_2 molecules with the active centers of hemoglobin. Since the immobilization of labels upon the addition of O_2 in native hemoglobin ($\alpha_2\beta_2$) is more intense than in the free chains, they concluded that the main allosteric effects are on the surface of the β subunits in the area of contact with the α subunits.

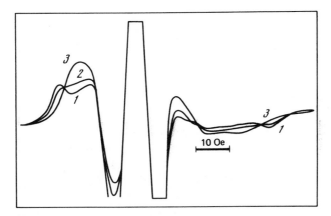

Fig. 40. Electron paramagnetic resonance spectra of radical IV on the β-93 SH groups of completely oxygenated (*1*), partially oxygenated (*2*), and completely deoxygenated (*3*) hemoglobin [1].

This conclusion was confirmed by a very ingenious method in another paper by McConnell et al. [240], who reconstructed hemoglobin by starting with carboxy (α-CO and β-CO), oxy (α-O_2 and β-O_2), and deoxy (α and β) derivatives of the subunits. Since the affinity of the heme group to CO is higher than to O_2, and the dissociation rate of the carboxy derivatives is low, the respective initial states of the subunits were fixed on assembly of the hemoglobin. It turned out that the hybrid structures α-O_2, β-CO, and α,β-CO with labels on the subunits produced practically identical EPR spectra, while the spectra of α-CO, β, and β-O_2, α-CO differ significantly from one another, the spectra of the latter corresponding to the spectrum of completely oxygenated hemoglobin (α-O_2, β-O_2).

A more detailed investigation of hemoglobin labeled with radicals IV and V revealed details of the conformational changes [97]. The experimental spectrum of the preparation is a superposition of the two signals A and B, which correspond to two states with correlation times (τ) approximately equal to 1.5×10^{-8} and 3×10^{-9} sec. The relationship between the signals is dependent on the external conditions (temperature, pH, ionic strength, and aggregation state). McConnell et al. explained the results obtained by the existence of two conformations. In conformation A the nitroxide tip of the label dislodges the β-145 tyrosine from the cavity of the protein and is thus strongly immobilized; in conformation B the label is outside of the cavity and rotates rapidly. Experiments with oriented single crystals have also clearly demonstrated the existence of two conformations. For example, a particular spectrum [97] displayed two signals with different hyperfine splitting.

Such a scheme is readily confirmed by X-ray diffraction analysis [216. Spin labeling slightly alters the structure of the protein in the vicinity of the labels added and decreases the cooperative properties (the value of the Hill coefficient drops from 2.9 to 2.3). The entry of molecular oxygen into the heme centers of the protein causes significant changes in the spectra, which indicates an increase in the fraction of the more immobilized component A. The B → A transition is effected by means of intermediate forms, and there is no isobectic point in the titration of the preparation with oxygen. Analogous effects were observed in reactions of deoxy forms of hemoglobin from various sources [241,242].

A more detailed investigation that included the use of hybrid hemoglobin forms in which the α and β fragments were in different, ferro, and ferri states, as well as the application of chemically and genetically modified preparations, made it possible to follow the competitive pathways for transmission of the allosteric effects from one subunit to another. For example, such changes in the hemoglobin structure as the splitting off of the β-146 histidine and the β-145 tyrosine by carboxypeptidase and replacement of the distal histidine by arginine result in a drop in the degree of cooperativity and the appearance of isobectic points. Analysis of these data led Baldassare et al. [242] to conclude that the amino acids on the boundary between the α and β subunits play an important role in the transmission of allosteric effects. These results are consistent with the conclusions of Perutz, who stated that the C end of the β chain and especially the β-145 tyrosine play an important role in the regulation of the electronic state of the heme group and in the realization of the Bohr effect.

In another series of experiments [243] the heme–heme interaction in hemoglobin was studied by introducing a spin label into the heme group. This was possible by modifying protohemin at one of the propionic groups and then introducing the heme into the α and β chains. Hybridization of the met form of the labeled subunits with deoxy forms of the unlabeled complementary hemoglobin fragments permitted Asakura and Drott to follow the state of the spin label in the vicinity of one of the heme groups as it acts on the other [243]. This ingenious technique revealed that there is a significant mutual effect between the states of the hemes in the α and β chains. Experiments with different ligands showed that dipole–dipole interactions between the paramagnetic centers on different subunits make a significant contribution (Fig. 41).

The use of spin labels XLV and XLVI, which are derivatives of ATP and triphosphate, in the regulatory centers of hemoglobin and mutant hemoglobin (leucine-12 instead of the α-92 arginine) made it possible for

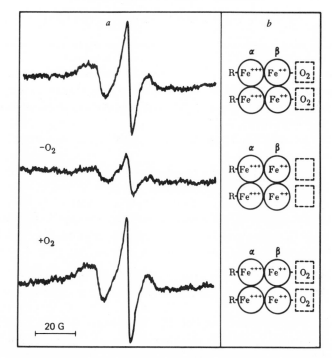

Figure 41. Allosteric effects in hemoglobin with heme groups labeled with nitroxide radicals at 25° [243]: (*a*) EPR spectra; (*b*) patterns of the addition of labels to the α-chains of hemoglobin and of the incorporation of O_2 in the heme groups of the β-chains; (*R*) the spin labels.

Ogata and McConnell [129,134] to monitor the interaction between the regulatory and acceptor centers of the protein. A thorough quantitative analysis of the curves describing the saturation of hemoglobin with carbon monoxide and radicals revealed a good fit between the experimental data and the results predicted by the generalized concerted cooperative model of Monod, Wyman, and Changeux.

The interesting prospects in the study of allosteric enzymes with spin labels have been demonstrated [244] with investigation of aspartate trans-carbamylase. The enzyme was modified with label XXX only in the absence of the substrate (succinate) and the allosteric activator (ATP), while some of the label occupied positions on the catalytic subunits. Addition of the allosteric inhibitor CTP and activator (ATP) had an effect on the EPR spectrum, but these changes did not correspond quantitatively to those predicted on the basis of a simple allosteric model of heterotropic effects.

GENERALIZED CONCERTED TRANSITIONS

The work of McConnell and his coworkers, however, has not completely resolved the question of allosteric interactions. Hemoglobin, the carrier of oxygen, is actually not an enzyme. The main dramatic events that occur in hemoglobin when it interacts with substrates take place on the boundary between the subunits and can be induced by disturbing the subunits without significant changes in their tertiary structure. It would be interesting to investigate conformational changes in enzymes that do not have a quaternary structure.

This is the reasoning behind a series of studies [4,5,30–33,81–83,113,114], whose purpose was to experimentally investigate allosteric effects in the enzymes lysozyme, myoglobin, aspartate aminotransferase, and myosin. Differing with respect to macromolecular size, enzymatic function, and other properties, these enzymes are convenient systems for studying the general laws governing enzyme action.

Lysozyme [3,83,113,114].   Here the study of the effect of specific inhibitors of lysozyme, particularly fragments of polysaccharide chains, is of special interest. According to the X-ray diffraction analysis of Phillips, the latter are embedded in the cleft in the vicinity of subsites B and C and thereby prevent the formation of enzyme-substrate complexes. According to Fig. 42, the addition of sucrose, glucose, and maltose, which are either not bound at all to the active center or are bound very weakly, to lysozyme preparations labeled at histidine-15 has little effect on the EPR spectrum of the label. However, the addition of much smaller quantities of the spe-

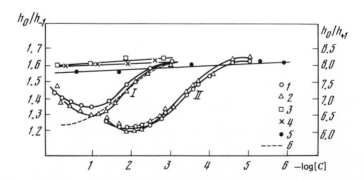

Figure 42.   Variation in the relative intensity of the EPR components of radical VII on the histidine-15 group of lysozyme as a function of the logarithm of the concentration of specific inhibitors of lysozyme [30]; (*1*) h_0/h_{+1}, NAG (I) and NAG–NAG (II); (*2*) h_0/h_{-1}, NAG (I) and NAG–NAG (II); (*3–5*) h_0/h_{-1}, glucose, sucrose, and maltose, respectively; (*6*) binding of NAG and NAG–NAG to active centers of lysozyme.

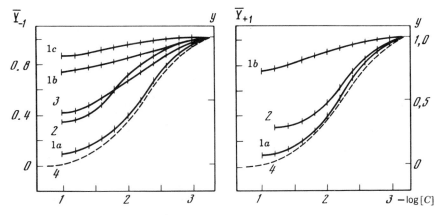

Figure 43. Variation in the relative EPR parameters h_0/h_{-1} and h_0/h_{+1} ($\overline{Y}_{\pm}$) and in the relative activity of lysozyme (y) as functions of the logarithm of the concentration of the inhibitor NAG. The value of $\overline{Y}_{\pm}$ is taken as unity for $[C] \to 0$ and as zero for $[C] \to \infty$; (1a–1c, 2,3) spin-labeled lysozyme preparations (see Table 1); (4) binding of lysozyme by the specific inhibitor NAG [113].

cific inhibitors NAG and NAG–NAG induces distinct changes in the EPR spectra. The relative changes in the value of the parameters h_0/h_{+1} and h_0/h_{-1} as functions of the concentration are in good quantitative agreement with the extent of binding with the substrates [245]. The maximum for NAG, added to subsite C of the active center, is about 25%, and that for NAG–NAG is about 40%.

The increase in the rotational diffusion parameter of the spin label possibly indicates that there is some separation between the protein segments in the vicinity of the histidine group on the opposite side of the protein globule at a distance of 20 Å from the active center when inhibitors are bound to the active center. According to X-ray diffraction data, NMR spectra [246], and differential fluorescence spectra [247], this binding causes some contraction of the groups in the active center. If we assume that most of the label is concentrated on the histidine group of lysozyme, we must admit that there is a specific cooperative transition extending from subsite C to the region of the histidine-15 group. The main results regarding the effect of the binding of the specific inhibitor NAG to lysozyme on the EPR spectrum parameters of spin labels added to various segments of the protein globule (see Chapter 2) are presented in Fig. 43 and Table 1.

These results may be summarized in the following manner:

1. In the case of almost all the spin-labeled lysozyme preparations listed in Table 1, there is a correlation between the changes in the parameters

of the EPR spectra and the extent of binding of the inhibitor to the
active center of the enzyme.

2. As the flexibility of the labels added to the his-15 group in lysozyme
preparations 1a, 1b, and 1c (Table 1) is increased, the relative changes
in the spectra $(Y_\pm)$ become weaker (see Fig. 43). Similarly, the effect is
weaker for lysozyme preparation 2 (lysine groups) than for the less
flexible "leg" in lysozyme preparation 1a. The weakening of the effect
seems to be due to the fact that by being spread from subsite C of the
active center to the opposite side through the protein globule it becomes
weaker as the binding with the matrix becomes less in the more flexible
labels.

3. Spin labels on the longer and more flexible lysine groups of lysozyme
preparations 2 and 5 are less sensitive to the implantation of NAG in
the active center than are the shorter labels on the his-15 group, but
are more sensitive than the less flexible labels. That is, the "conformational wave" is not only transmitted into the vicinity of the his-15
group, but is also directed with even greater intensity toward the lysine
groups.

4. The maleimide label on his-15 and probably on the terminal α-NH_2
group of lysine (lysozyme 3), which is characterized by a fairly high-
frequency ν, is more sensitive to the effect of an inhibitor than are the
shorter labels on his-15 in lysozymes 1a, 1b, and 1c. We may assume
that the conformational wave reaches the α-NH_2 terminal group on lys-1.

5. Spin label III, when added to the asp-52 group (lysozyme 4), is practically insensitive to the addition of NAG, possibly because of a certain
alteration in the structure of the enzyme on modification.

6. In the case of lysozyme 1a with NAG–NAG in subsites B and C of the
active center, the addition of the inhibitor has a stronger effect on the
EPR spectra than does the binding of the inhibitor only to subsite
C [30].

The data obtained indicate that the conformational waves that cause
the changes in the state of the labels in the vicinity of the his-15 group are
most likely propagated over the more superficial layers of the protein
matrix, possibly through a broad cleft along the path C → B → A → lys-
96-97 → his-15 rather than along the direct route NAG acetylamide
group → tryptophan-108 → internal groups → histidine-15 from subsite C.

This scheme is only preliminary. However, it very reliably demonstrates
that, when a specific inhibitor is bound to subsite C, the effect of the conformational changes is not localized in any one place in the macromolecule,
but spreads into different parts of the protein globule, reaching distances
of at least 15–17 Å. It must be noted that the level of the conformational

changes is very low, but they have been detected in many portions of the lysozyme protein macromolecule, and their overall thermodynamic contribution can be significant. The conclusion that conformational effects are significant factors in the thermodynamic parameters for the binding of NAG and its oligomers has been drawn [249] on the basis of calorimetric measurements.

The results cited in Chapter 7 suggest that the state of the water has a great effect on the effective rotation of spin labels on the relatively independent lysine groups and of comparatively long labels on the his-15 group. Presumably, the constriction of the nearby functional groups by substrates brings the two tightly packed polypeptide nuclei of the macromolecule closer to one another and indirectly induces a change in the state of the water, which is detected by the spin labels.

Myosin. Incubation of rabbit myosin with radicals I and IV (see p. 62) results in the addition of the latter to the sulfhydryl groups in an amount equivalent to 1–17 labels per protein macromolecule, depending on the nature and concentration of the reagents [31,81,155,250,251].

The addition of the first labels causes a four- to sixfold increase in ATPase activity. Radical IV first selectively blocks sulfhydryl group SH I, whose modification results in activation of the enzyme [252]. This conclusion is confirmed by the fact that the activity of the enzyme is independent of further additions of radical V. The effect of radical I is less specific, since the maximum activation of the enzyme is observed when approximately five spin labels are added. A further increase in the number of radicals bound to the enzyme (≤ 11 per protein molecule) results in a loss of enzymatic activity, which is apparently due to the blocking of the SH II group of the active center [252]. The EPR spectrum of spin label IV, which is preferentially bound to sulfhydryl group SH I, indicates that the nitroxide fragment is greatly immobilized and that there are at least two binding centers with different conformations of the local environment [31]. The selective blocking of the SH I group is due to its high chemical activity, despite the fact that, according to the EPR spectrum, it is sterically less accessible than the other SH groups.

It is known [253] that myosin forms a ternary complex with ATP and several divalent metal ions (Mg^{2+}, Ca^{2+}, and Mn^{2+}). The effect of these components on the EPR spectra of spin labels IV and I, when covalently and preferentially bound to the activating sulfhydryl group SH I, has been studied [31]. According to Fig. 44, the action of Ca^{2+}-ATP, Mg^{2+}-ATP, and Mn^{2+}-ATP substrate mixtures on samples of spin-labeled myosin causes appreciable changes in the nature of the spectra. These changes do not appear when the metal ions are added without ATP. Immediately after the addition of the substrate mixtures, the ratio of the intensities of the spectral components for myosin labeled with both radicals varies by 15–40%, depending on the type of label, the nature of the metal ion, and the ATP concentration with an average measurement accuracy of 5%. The nature of the effect of the paramagnetic Mn^{2+} ion does not differ, in principle, from that of diamagnetic ions. It appears that the activating SH group is separated from the active center by a distance that does not allow a contact or a strong dipole–dipole interaction between the nitroxide labels and the paramagnetic manganese ion (see Chapter 3).

Further incubation of spin-labeled myosin samples with a substrate for 5–10 min results in the gradual disappearance of the EPR spectral component corresponding to the strongly immobilized labels IV and I. As seen in Fig. 44, the decrease with time in the intensity of the strongly immobilized component of the EPR spectrum is accompanied by a symbatic increase in the intensity of the less immobilized component. Thus when myosin interacts with substrates, there are apparently two local conformational changes due to the changes in the state of the labels. The first change occurs immediately after the substrates act, and the second occurs over a long period of time and is most likely due to the formation of a complex of the enzyme with ADP.

Consideration of Fig. 44 reveals that in all cases the formation of a myosin–metal–ATP ternary complex results in a decrease in the ratio of the EPR spectrum components; that is, the radical becomes less immobilized, indicating some loosening in the protein structure. This result is in agreement with the kinetic data for the myosin–Ca^{2+}–ATP system, for which it has been shown that the formation of the enzyme substrate com-

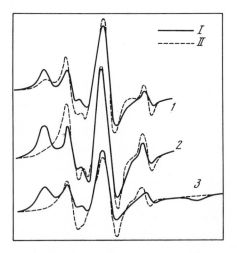

Figure 44. Variation in the conformation of myosin in the vicinity of the SH I group during the formation of the ternary complex with ATP and metal ions [31]: (*I*) myosin (10 mg/ml) with spin label IV; (*II*) myosin (10 mg/ml) with spin label IV on addition of a mixture of ATP (2.5×10^{-4} M) with $CaCl_2$ (2.5×10^{-4} M) (1); $MnCl_2$ (7.5×10^{-4} M) (2); $MgCl_2$ (7×10^{-3} M) (3).

plex is accompanied by loosening of the protein structure [254,255]. Moreover the optical rotatory dispersion of myosin remains practically unchanged on addition of ATP and ADP, indicating the absence of large-scale conformational changes in the protein globule [253]. Presumably, the change in the conformation of the protein matrix is local. Similar results were obtained by other investigators [102,147–150], who found that modification of the SH I group with labels I and IV causes activation of the enzyme. Various effects on the process, such as the formation of enzyme-substrate (ATP) and enzyme-inhibitor (ADP and diphosphate) complexes, the blocking of other SH groups, temperature increases, and so forth, are accompanied by a decrease in the immobilization of the spin labels on the SH I group, the addition of ATP causing a greater conformational change than the addition of ADP and other derivatives. When actin acts on spin-labeled myosin, the spectrum corresponds to an increase in the fraction of the more immobilized state; that is, the structure of the enzyme in the vicinity of the SH I group becomes more compact [256].

Other enzymes and proteins. Sperm-whale myoglobin is a comparatively low-molecular and compact protein with a known structure. The three easily accessible myoglobin histidine groups CD-6, EF-4, and CH-1 were modified with labels IV, VI, and VIII [32,81,83]. According to the Kendrew model and the data obtained with the aid of double labels, each of these groups is separated from the heme center by a distance of about 16–18 Å and is thus inaccessible to the direct action of the substrates at the heme active center. According to Fig. 45, when myoglobin is titrated with the specific inhibitor cyanide, the curve describing the relative change in the

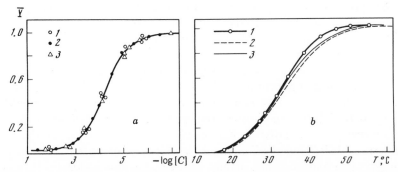

Figure 45. Variation in the relative parameters of the EPR spectra of myoglobin ($\overline{Y}$) spin labeled at the accessible histidine groups during complexation with a specific inhibitor (cyanide) (a) and on increasing temperature (b) [32]; (a) (1) h_0/h_{+1}; (2) h_0/h_{-1}; (3) binding of cyanide; (b) (1) relative EPR spectral parameters; (2) specific heat; (3) rate of proton relaxation.

EPR spectral parameters of label VII practically coincides with the curve describing the binding of cyanide to the active center. This result indicates that the conformational changes in the enzyme resulting from changes in the electronic state of the heme group are cooperative.

Highly purified preparations of swine-heart aspartate aminotransferase (AAT) were labeled with radicals IV and VIII at the readily accessible SH groups with amounts equivalent to one or two labels per enzyme subunit (the molecular weight of the subunits is 45,000). The modification has practically no effect on the enzymatic activity or on the absorption spectra of the pyridoxal–phosphate center of AAT. Introduction of a copper atom into the active center of the enzyme has no effect on the EPR spectrum shape for frozen solutions. According to the data in Chapter 3, this indicates that the label is separated from the substrate-binding site by a considerable distance ($>$16–18 Å). Saturation of AAT with specific substrates containing an anionic carboxyl group causes reproducible changes in the EPR spectra of the spin labels (Fig. 46). Apparently, the interaction between the anionic substrate group and the nearby cationic center of the enzyme is transmitted over the protein to distances greater than 18 Å.

In [115,143] a steroid–serum albumin system was used to analyze the possibility of the transmission of a conformational wave over a protein matrix in the case of a nonspecific interaction, as in the case of generalized concerted transitions in enzymes [3]. To this end, an investigation was carried out with the aim of ascertaining the conformational changes in serum albumin during interactions with steroids and other compounds. The EPR spectrum of serum albumin labeled with radical VII is the sum of two signals: (a) signal A, which fits a strongly immobilized nitroxide radical with a correlation time (τ_c) approximately equal to 10^{-8} sec and (b) signal B, which fits a weakly immobilized radical with $\tau_c = 5 \times 10^{-10}$ sec.

The presence of the steroids prednisolone, 6-methylprednisolone, hydrocortisone hemisuccinate, Viadril, methylandrostanol, estradiol, methandrostenolone, and indomethacin causes considerable changes in the EPR spectra of labeled albumin. The most significant effect is the increase in the intensity of component B, which indicates an increase in the rate of rotational diffusion and an increase in the fraction of radicals responsible for signal B. Diluting the protein preparations to a concentration of 1–2 mg/ml does not qualitatively alter the effect of indomethacin on the state of the spin labels. Thus the binding of steroids is accompanied by loosening of the protein matrix in the vicinity of the spin labels added.

The results of experiments involving introduction of a hydrophobic spin probe into the binding portion of the albumin macromolecule show that the probe molecules and consequently the steroid hormones displacing them are located at distances greater than 16–18 Å from spin labels I and

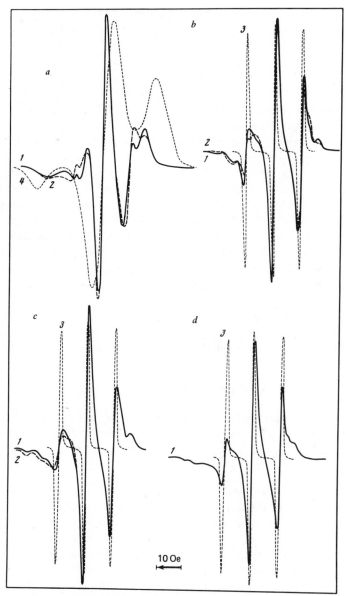

Figure 46. Electron paramagnetic resonance spectra of radicals VIII (*a*), VI (*b*), IV (*c*), and I (*d*) covalently bonded to aspartate aminotransferase [34]: (*1*) without ligands; (*2*) with ligands; (*3*) in water; (*4*) in a water–ethylene glycol mixture (at 77°K).

VII (see Chapter 3). This excludes the possibility of a direct (isoteric) interaction. We cannot completely rule out the possibility that the phenomena observed in the experiments with concentrated protein solutions (15–20 mg/ml) are due to dissociation of the albumin aggregates on binding of the steroids. However, the conformational changes in serum albumin in the experiments with the diluted solutions apparently occur by an allosteric mechanism, as in the case of the cooperative changes observed when specific effects act on enzymes [1–3].

The complexing of donkey antibodies with human γ globulin is accompanied by conformational changes in both proteins [257]. This is indicated by changes in the spectra of spin label VII bound to these proteins. The experiments were carried out in such a manner that in some cases only the antibodies were labeled, while in other cases only the γ globulin was labeled. The most noticeable effect from the addition of the complementary protein was the change in the fraction of signals from the immobilized and freer states of the spin label. A very interesting result of these experiments was the proof that the complexation has different effects on the conformational states of the two proteins. The spin labels on the γ globulin were immobilized, while those on the antibodies experienced a reduction in immobilization. These data seem to demonstrate the general ability of protein matrices to contract in one portion while expanding in another. Such changes are not transmissible from one macromolecule to another.

ACTIVE CENTERS OF ENZYMES

Spin-label techniques have created interesting possibilities for investigating the topography and dynamic structure of active centers in enzymes. Although these possibilities have still not been realized to an adequate extent, new results are now being obtained in experiments with many enzymes.

Serine proteases. Chymotrypsin.   The first attempt to study the active center of an enzyme with the aid of spin labels was undertaken by Berliner and McConnell [104], who modified the ser-195 group of α-chymotrypsin at pH 3. The spectrum of label XII indicates that there is fairly strong immobilization of the nitroxide fragment due to the inclusion of the label in the comparatively narrow cleft of the active center. Bringing the ambient pH to a neutral value results in deacylation of the enzyme. Rotation of a single crystal in a magnetic field revealed the existence of a well-defined orientation of the nitroxide fragment relative to the protein macromolecule [258]. A more careful look at the data of these workers leads to the conclusion that there is a possibility of several more random orientations, all of which amount to half of the total, in the samples investigated.

Next a whole series of reagents for the active serine group in proteolytic enzymes based on phosphate and nitrobenzene derivatives was proposed [65,104–108,137]. Using these derivatives, researchers were able to compare the active centers of various proteases with the active center of α-chymotrypsin, whose structure is well known. The most conclusive work in this area was carried out by Piette, Kosman, and Hsia [105,259,260], who were the first to use the biradical label XIV. The EPR spectrum of spin-labeled cholinesterase demonstrates that there is much greater freedom in the motion of the biradical fragment than in the α-chymotrypsin. The active center of cholinesterase has a more open structure. This approach was extended to a broader range of serine proteases. The enzymes investigated can be arranged in the following sequence according to the degree of immobilization of the biradical spin label XIV: α-chymotrypsin $>$ subtilisin Carlsberg $>$ trypsin $>$ elastase $\sim$ thrombin $\sim$ subtilisin BPN' $\sim$ bovine erythrocyte cholinesterase $>$ eel cholinesterase $>$ horse-serum butyryl cholinesterase. The kinetics of the addition of the spin label to the β-195 serine group of α-chymotrypsin is pH-dependent. The optimum in the plot corresponds to the state of the enzyme with a deprotonated histidine-57 group and a protonated isoleucine-16 group. The EPR spectra of all biradical spin-labeled enzymes showed five lines, indicating that the nitroxide fragments are forced to approach one another as a result of the action of the walls of the active centers (see Chapter 3). This fact is manifested in particular by the fact that unfolding the protein with urea (8 M) and guanidine (6 M) transforms the spectrum of the spin-labeled radical into a triplet characteristic of weakly interacting nitroxide fragments separated by a considerable distance.

A series of *ortho*-, *meta*-, and *para*-substituted benzylsulfonyl fluorides (inhibitors of α-chymotrypsin and trypsin) was employed in a comparative study of the active centers of these enzymes. It turned out that the *meta* derivatives are the most sensitive to the structure of the enzyme and the labels [137]. The great prospects in the combined application of labels and chemical modification of enzymes was demonstrated by Kosman et al. [109], whose experiments involved the successive addition of spin labels XVIII–XX of approximately equal length to the ser-195 group of α-chymotrypsin. The mobilities of the labels do not correlate with their flexibilities. For example, the less flexible label XVIII is less immobilized than the more flexible label XIX. This indicates that there is a specific interaction between the labels with the aryl (met-192) and amide (met-180) segments of the active center of the enzyme. The nitroxide fragment of label XX rotates fairly rapidly and, thus, does not come in contact with the protein matrix. It was found that indole has different effects on labels XVIII and XIX bound to the active center of chymotrypsin and on the rate of deacylation of spin-labeled acyl enzymes. Thus, the rotation of label XIX under the effect of indole remains free, and the deacylation rate of the acyl enzyme increases, while in the case of spin label XVIII the reverse is true, with the motion of the label on the protein slowed and the rate of deacylation dropping considerably. This means that the binding of indole in the aryl center causes a change in the amide segment. This result indicates that there are cooperative interactions between two different portions of the active center of chymotrypsin. Methionine-192 appears to play an important role in this cooperative interaction, since oxidation of methionine-192 does not result in changes in the EPR spectra of labels XVIII–XX under the effect of indole.

These workers stated on the basis of the experimental results obtained that the steps in the process of enzymatic hydrolysis occur in the following order: (a) the substrate and the enzyme interact first at the aryl- and amide-binding centers, (b) the binding in the aryl segment induces conformational changes in the enzyme that orient the amide-binding segment and the catalytic center, and (c) bonds in the aryl segment are cleaved as a result of acylation. The deacylation, which completes the sequence of steps, occurs in the looser conformation. Subsequent experiments [259–260] revealed the extent of alteration in the microstructure of the active center of α-chymotrypsin when various portions of it were subjected to specific chemical modification. Oxidation and alkylation of the met-192 and met-180 groups causes freer motion of the biradical label on the ser-195 residue in the following order: the sulfoxide of met-192 < the disulfoxide of met-192, 180 < carboxymethylated met-192 < carboxymethylated met-180, and 192 < the sulfoxide of met-192 and carboxymethylated met-180. These laws, of course, cannot be explained purely on the basis of steric effects, which should have caused the nitroxide fragments of the biradical labels to approach one another. Thus modification of the methionine groups of the enzyme disturbs the local structure and opens up the cleft of the active center. A similar but weaker effect is caused by the blocking of the catalytic serine group by dicyclohexylfluorophosphate. Here spin label XIV bound to the met-192 group becomes somewhat less immobilized.

It has been hypothesized [111,116,261] that the isoleucine-16 group of α-chymotrypsin controls the conformation of the active center. The effect of the degree of ionization of this group on the conformation of the enzyme in the vicinity of his-57 and met-192 has been thoroughly studied [111]. The EPR spectra of the spin labels respond differently to an increase in pH from 7 to 11. In the case of spin-labeled met-192 (label IV), there is an increase in the rotation frequency with increasing pH. The presence of an isobectic point indicates that at alkaline pH there are two enzyme conformations, the transition between which is associated with the ionization of isoleucine-16 (pK_a = 8.9). Indole has practically no effect of the EPR spectrum of label XXX at neutral pH, but in alkaline media (pH > 7), the binding of indole alters the dependence of the spectral characteristics on the pH, the value of pK_a for isoleucine-16 increasing to 10.1. Spin label XXV bound to the his-57 group of the chymotrypsin-active center inhibits the activity of the enzyme. When the pH is increased from 7 to 11, the motion of the label is slowed because of the conformational change between the two forms of the enzyme (pK_a = 8.9). Unlike the preceding case, at neutral pH, indole and other reagents added to the spin-labeled enzyme cause an increase in the rotation frequency of the label, and the pK_a of isoleucine-16 increases to 10.1.

Thus different portions of the active center of α-chymotrypsin react in qualitatively different manners to changes in the charge on the isoleucine-16 group and to the entry of indole into the hydrophobic binding center of the enzyme. The experiments with spin-labeled chymotrypsin and other proteinases can thus serve as examples for the study of the fine structure and conformational changes in the active centers of enzymes with the aid of spin labels.

Actin. In the preceding section of this chapter we described the results of investigations on the local topography of the active center of myosin and the effect of specific substrates of the enzyme on its conformation in the vicinity of a spin label on the SH I group. The modification of the SH groups of actin (1 mole per molecular weight of 46,000) has been the subject of several reports [102,151,152,256]. Maleimide derivatives of a five-membered nitroxide radical of various lengths were employed. The number of bonds (n) between the chemically active group and the radical fragment, which allow free rotation or rotameric transitions, varied regularly from one to eight. The level of unhindered motion of the labels increased along the same series. The most abrupt changes in the spectra were observed in the transition from n = 5 to n = 6. The length of the label with n = 5 apparently fits the length of the trough at the bottom of which the sulfhydryl group is found. A similar result was also obtained in experiments on the orientation of F-actin films, which has a strong influence on the EPR spectra of the shortest label, a weaker influence on the labels with n = 2 and n = 5, and practically no effect on radicals with n > 5. The interaction between actin and an ATP molecule with a nitroxide fragment added to its adenine group has been studied [121]. The label was bound to the enzyme molecule in such a manner that the fragment was slowed to a rotation rate similar to that of the protein macromolecule. The ATP adenine group was thus tightly held in the active center of the enzyme.

The polymerization of actin is clearly monitored with the aid of spin labels on the SH groups of the enzyme [262]. In order to evaluate the distance r between a maleimide spin label on the SH groups of actin and the segment for the binding of divalent metals, Gergely et al. [263] used the method developed by Taylor et al. [21]. Introduction of a paramagnetic Mn^{2+} reduces the intensity of the spectrum of the label by about 50% compared to the spectrum in the presence of the diamagnetic ions Ca^{2+} and Mg^{2+}. As a result the value of r equal to 20 ± 3 Å was obtained. The polymerization of actin does not produce any significant change in this distance.

The conformational changes of spin-labeled actin in the actin–troponin–tropomyosin complex were observed by Tonomura et al. [264], who demonstrated a change in the state of the spin label on the tropomyosin in the presence of troponin under the effect of the Ca^{2+} ion.

Dehydrogenases [103,117–120]. The most thoroughly studied enzyme of this group is liver alcohol dehydrogenase. Spin labels XXXV–XXXVII have been introduced into different portions of the active center of the enzyme, specifically into the NAD- and Zn-binding centers and into the SH group.

The variation in the relaxation rate of water and ethanol protons ($1/T_1$ and $1/T_2$, where T_1 and T_2 are the longitudinal and transverse relaxation times of protons, respectively) yields additional information on the structure and properties of the active center and the enzyme-substrate complex. In particular, it has been shown that: (a) label XXXVI reversibly and competitively displaces NAD, (b) spin labels on the SH group and the Zn-binding site are located in direct proximity to one another, and (c) the nitroxide fragment of the label replacing the nicotinamide ring of NAD is readily accessible to water and ethanol molecules [117–120].

Careful measurement of the NMR spectra of ethanol, acetaldehyde, and other specific reagents in the presence of the enzyme-label complex permitted Mildvan and Weiner [119] to evaluate the magnitude of the electron-nuclear dipolar interaction and to then calculate the distance from the nitroxide fragment in the active center to the various protons in the substrate (for further details, see Chapter 5).

The SH group in the active center of D-glyceraldehyde-3-phosphate dehydrogenase was labeled with derivatives of XI and XVII with different values of n. An increase in n is accompanied by an increase in the fraction of the less immobilized component and a rise in the mobility. The binding of NAD results in an increase in the freedom of motion of the labels, that is, loosening of the protein matrix in the vicinity of the SH group [103].

Other proteins. Enzymes and carriers. A new approach to the comparative study of the topography of the active centers of hemoproteins was devised by Asacura et al. [265]. The two propionic residues of a heme group were modified with spin labels, and the heme group was then introduced in the active centers of myoglobin, hemoglobin, horseradish peroxidase, and cytochrome C peroxidase. It turned out that all of these substances differed with respect to the magnitude of the spin–spin (radical–radical and radical–heme) interaction. The features of the structure of the apo-enzyme in the vicinity of the active centers govern the specific nature of the conformations of the labeled heme groups. In hemoglobin and myoglobin one nitroxide fragment is pressed onto the heme, and the other remains free. In the cytochrome peroxidase both fragments are free, but in the horseradish peroxidase both are immobilized.

The possibility of determining the depth of the cleft of an active center with the aid of a set of spin labels with different lengths was first demonstrated by Hsia and Piette [136]. Increasing the chain length of radicals introduced by means of a dinitrophenyl fragment into the active centers of antibodies begins to have an effect on the EPR spectrum when this length reaches a certain critical value. A set of *ortho*- and *para*-nitrobenzene derivatives of spin-labeled haptens has been used to study the heterogeneity of the active centers of antibodies [266].

Ribonuclease A, presumably labeled at the his-12 and his-119 groups (compounds XXII–XXV) and the lysine group (VIII) of the active center, does not exhibit enzymatic activity. Nevertheless the addition of RNA, polyadenylic acid, and polycytidylic acid causes considerable immobilization of the spin labels. This indicates that the substrate and inhibitor-binding centers are outside of the catalytic active center [110].

The position of spin label XLIV, a phosphate derivative, in the active center of ribonuclease was studied with the aid of NMR according to the broadening of the lines for the protons of his-12 and his-119 [128].

A set of spin-labeled sulfonamide inhibitors of carbonic anhydrases has been used for a comparative analysis of the structure of enzymes from various sources [125–126]. It was shown, in particular, that the active center C of the enzyme is a narrow hydrophobic cleft approximately 14.5 Å deep. Active center B is similar in structure to center C, but is located in a narrower cleft. A higher level of radical mobility was observed in the labeled preparations of human carbonic anhydrase than in the preparations isolated from bulls.

The synthesis of an isocyanide derivative of radical XLVII has opened up some interesting possibilities for introducing spin labels into the active centers of heme enzymes. The changes in the rotational diffusion of a spin label based on a six-membered derivative and joined by its isocyanide residue to a heme group at various pH values correlate with the changes in the oxidation–reduction potential of cytochrome C. The radical has also been introduced into the active centers of cytochrome P_{450} [130–133].

The foregoing experimental material indicates that there are fairly abundant possibilities for spin-labeling methods in the study of the active centers of enzymes of various classes. The parameters of the EPR spectra of spin labels reflect the topography of active centers to some extent. In a number of cases it has been possible to reveal conformational changes in different portions of a center as well as the extent to which chemical modification of specific groups in enzymes have an effect.

In the case of the addition of labels to active centers, as in the case of any chemical modification, there is a drop in enzymatic activity and possible local conformational changes. Nevertheless the experiments on enzymes and especially on α-chymotrypsin showed that the unmodified adjacent portions of the active centers and the adjacent protein matrix can maintain their functions. The reliability of the information obtained can be increased by using a whole set of labels and comparing the results with data from other methods.

Chapter Seven
Local Mobility of Water–protein Matrices of Proteins and Enzymes

The solution of various problems in enzymology (Chapter 9) requires data on the local viscosity and local mobility of various portions of water–protein matrices of enzymes. The mobility of a macromolecule is related to such important characteristics as molecular weight, size, and shape. The comparatively rapid method of determining the rates of rotational and translational diffusion of nitroxide radicals from the parameters of their EPR spectra [36–41] has opened up interesting prospects for the measurement of the rotation of biological macromolecules in solution and for the study of the local mobility of individual sections of biological structures and the surrounding aqueous shell. The possible uses of spin labels and probes are due primarily to the fact that, according to the contemporary theories, the elementary acts of diffusion in liquids and amorphous bodies occur at the moment a free volume is created due to the displacement of the surrounding particles. There are many arguments favoring the idea that the rotational diffusion parameters of nitroxide radicals reflect the state and mobility of the surrounding particles [3–5,25,36–38,50–53,60,79].

This leads to the proposal of a complex approach to the study of the local mobility of specific sections of water–protein matrices based on the following experiments: (a) measurement of the temperature dependence of the rotational diffusion of spin labels of various lengths and flexibilities, (b) determination of their availability for encounters with paramagnetic probes freely diffusing in the solution (see Fig. 28), (c) study of the mobility of radicals interacting with the more flexible layers of the protein matrix (e.g., hydrophobic probes), (d) comparison of the results obtained on proteins (hemoglobin, lysozyme, α-chymotrypsin, and albumin) with the data on model systems, (e) comparative analysis of the possible behavior of spin labels on proteins with precisely known structures and of their behavior in actual systems (lysozyme, myoglobin, and hemoglobin), and (f) investigation of the rotational diffusion of spin labels on proteins in media with different viscosity and in different aggregation states (solutions, suspensions, and dry preparations). These investigations, together with the known data on the hydrodynamic properties of proteins in solutions and the results from hydrogen exchange, make it possible to construct a preliminary scheme of local molecular motions in proteins. The introduction of labels and probes into the active centers of enzymes (α-chymotrypsin, ribonuclease A, and myoglobin) make it possible to obtain information on the local mobility of the medium in the areas where enzymatic reactions take place.

EXPERIMENTAL BASIS OF THE METHOD OF MODEL SYSTEMS

The study of the effect of the structure of the local environment on the rotational diffusion of nitroxide radicals has been the subject of a number of papers [25,36–38,52,53,267–288]. The following parameters were usually employed: (a) the rotational diffusion correlation time (τ_c) (see Chapter 1), (b) the rotation frequency of the radical ($\nu = 1/\tau$), (c) the rotational anisotropy (ε), (d) the effective energy (E_{eff}), (e) or the enthalpy ($\Delta H_{eff}^{\neq}$) and entropy ($\Delta S_{eff}^{\neq}$) of activation, which is defined by an equation expressing the dependence of ν on the temperature

$$\nu = \frac{kT}{h} \exp \frac{\Delta S_{eff}^{\neq}}{RT} \exp \frac{-\Delta H_{eff}^{\neq}}{RT}. \tag{7.1}$$

In various studies [3,32,81–83,113] the following additional parameters were used: the rate constant for paramagnetic relaxation of nitroxide radicals during encounters with external spin probes k, as well as the quantities E^{+}, E^{-}, ΔS^{+}, and ΔS^{-}, which reflect the dependence of the parameters ν_{+} and ν_{-} on the temperature. The latter are calculated from Eqs. (1.3) and (1.4). As shown in Chapter 1, the differences in these parameters are due to the anisotropy in the rotation of the nitroxide radicals. Below we present the principal results of quantitative studies on the behavior of radicals in various media.

Many systems display a functional relationship between τ and the viscosity of the solution η that follows from the Stokes–Einstein equation (radicals in different solvents and at different pressures) [62,281]. The values of E_{eff} in pure liquids and water–glycerol mixtures are approximately equal to the values of the activation energies for flow in these systems [267]. In the case of many polymers, there is a satisfactory quantitative correlation between E_{eff} and $\Delta E_{eff}^{\neq}$ for rotational diffusion and the corresponding activation parameters for segmental motion, determined by independent methods [268,269] (Table 8). The values of the activation energy and the pre-exponential factors for nitroxide radicals in low-viscosity liquids lie within the limits expected on the basis of the kinetic theory of liquids [280] (see below).

Table 8. Activation Energies (E) and Preexponential Factors (ν_0) of the Rotation Frequencies of Radical Probe R and of Segmental Motion in Polymers (According to NMR Data) [268,270]

Solvent	$\nu_0 \cdot 10^{-12}$, sec^{-1}	E, kcal/mole	Solvent	$\nu_0 \cdot 10^{-12}$, sec^{-1}	E, kcal/mole
Ethylbenzene	11.2	3.3	Chlorobenzene	3.9	3.0
Toluene	2.45	2.5	m-Xylene	13.0	3.0
Cumene	36.0	4.5	Cyclohexane	6·3	3.1

Polymer	Probes		NMR	
	$E \pm 0.5$, kcal/mole	ν_0, sec^{-1}	$E \pm 0.5$, kcal/mole	ν_0, sec^{-1}
Isotactic polypropylene	10.5	9.0×10^{15}	8.2	1.0×10^{13}
Atactic polypropylene	18.7	2.2×10^{21}	17.0	2.0×10^{19}
Divinyl rubber	5.8	1.6×10^{14}	6.5	1.0×10^{13}
Polystyrene	18.2	4.0×10^{18}	—	—
Natural rubber[a]	8.5	3.4×10^{16}	10.5	1.0×10^{15}
	8.5	9.1×10^{15}	—	—
	8.5	1.7×10^{15}	—	—

[a]Probes of various sizes.

An increase in the rigidity of a polymer, for instance, by cross-linking its macromolecules with rigid bonds, is first accompanied by a parallel increase in E_{eff} and $\Delta S^{\neq}_{\text{eff}}$ and then by a decrease in these parameters when a certain limiting rigidity is achieved [268]. The structure of very rigid matrices appears to vary only slightly on heating, and a probe in a micro-cavity of the polymer undergoes rotational diffusion by overcoming the barrier of "friction" against the wall of the matrix. As the polymer is softened, its structure becomes increasingly sensitive to heating. The walls of the matrix acquire mobility, and the motion of the probe begins to re-flect the motion of the matrix.

Experimental data on the rotational diffusion of nitroxide radicals can be divided into groups [3,25]. Low-viscosity media like water are charac-terized by low values of E_{eff} and near-zero values of $\Delta S^{\neq}_{\text{eff}}$ [280]. In the case of more viscous systems like glycerol and its aqueous solutions [267], the increase in the viscosity of the system causes a simultaneous increase in E_{eff} and $\Delta S^{\neq}_{\text{eff}}$, there being a linear correlation between them (compensation effect), (i.e., $E_{eff} = 3.5 + 0.32 \; \Delta S_{eff}^{\neq}$ kcal). This group of media includes the soft, uncross-linked polymers [268], which obey the equation $E_{eff} = 3 + 0.32 \; \Delta S_{eff}$ kcal. Rigid systems like cross-linked polymers are characterized by low values of E_{eff} and negative values of $\Delta S_{eff}^{\neq}$ [268] (see below).

System	E_{eff}, kcal	$\Delta Sh_{eff}^{\neq}$, e.u.
Low–viscosity liquids (methanol)	2.5– 6.0	0–4
Moderately viscous liquids (water– glycerol mixtures)	5.0–11.0	2–16
Soft polymer systems with few cross-links	5.0–22.0	0–31
Rigid polymer systems with many cross-links	2.6–2.8	−5–0

A more detailed investigation [270] revealed the effect of the structure of the radical and the media on the parameters being measured. It turns out that the size and chemical structure of the probe has practically no effect on the value of E_{eff} for a particular polymer, while the value of $\Delta S_{eff}^{\neq}$ decreases within 5 e.u. with increasing probe size. However, E_{eff} is substantially dependent on the shape and state of the polymer. Different probes have different values for the anisotropy parameter ε. These values vary within one order of magnitude for polymers and by a factor of 2 for viscous liquids.

Thus there is some basis for a quantitative evaluation of the rigidity and local mobility of water–protein matrices of enzymes with the aid of spin labels. However, the binding of labels to protein macromolecules introduces some additional complicating factors due to the rotation of the macromolecule, the possibility of segmental motion in the "legs" of the labels, anisotropy of the local environment, and so on.

In view of these factors, we shall consider the possible variants of label motion and the respective changes in the experimental parameters (ν, E_{eff}, $\Delta S_{eff}^{\neq}$, ε, and k) as functions of the viscosity of the medium and the structure of the label [25] (Table 9).

As seen in Table 9, the study of spin-labeled protein preparations makes it possible to make a selection among various variants of the label–matrix interaction and to obtain information on the microstructure and local mobility of various layers of a matrix. Specific examples of the application of this technique will be given in the following sections.

Hemoglobin. The β-93 sulfhydryl group of bovine hemoglobin has been modified by spin labels I and IV–VIII, which have different lengths and

Table 9. Variants of the Interaction between a Spin Label and a Protein Matrix and the Respective Experimental Parameters (ν, E_{eff}, $\Delta S_{eff}^{\neq}$, and $k_w/2k_p$)

Variant	E_{eff}, kcal	$\Delta S_{eff}^{\neq}$, e.u.	ν, sec^{-1}	$k_w/2k_p{}^a$	Remarks
The label as a whole is rigidly packed into a rigid matrix and rotates along with it	4.4[b]	-5 to 10	$\sim10^7-10^8$	~1	Isotropic rotation according to Stokes's law
Part of the leg of the label is pressed into a rigid matrix, and part rotates freely in an aqueous medium	4.4	0 to -25	$10^{10}-10^7$	~1	The rotation is anisotropic, and E_{eff} is independent of the length. The dependence of τ on η is more abrupt than according to Stokes's law. The greater the immobilized part, the lower is $\Delta S_{eff}^{\neq}$
The label is absorbed in a comparatively soft protein matrix	4.4 to 20	5 to 50	$10^{10}-10^8$	<1	Anisotropic rotation of the nitroxide tip due the rigidity gradient of the matrix. The dependence of τ on the viscosity is weaker than according to Stokes's law.
The nitroxide tip of the label is packed into the hydrophobic part of the protein	<4.4	<0	?	$\ll1$	The hydrophobic bonds are strengthened with increasing temperature
The label is tightly packed in a portion of the matrix possessing local mobility relative to the macromolecule	?	?	?		τ is independent of the viscosity of the medium

[a] k_w and k_p are the rate constants for exchange relaxation of the nitroxide fragment of the spin label before and after addition to the protein. The coefficient 2 is introduced to take into account the slowing of the diffusion rate following addition to the protein.

[b] Activation energy for flow of pure water.

129

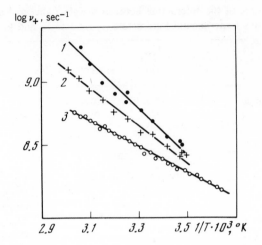

Figure 47. Arrhenius plots of the rotational diffusion parameter of spin label VI (see p. 24) on the β-93 SH group of bovine hemoglobin at various pH values [25,81,82]: (*1*) pH 5.0; (*2*) pH 6.0; (*3*) pH 7.0.

flexibilities, and the parameters E^+, E^-, ΔS^+, and ΔS^- have been measured for each of them (Fig. 47) [25,81,82].

Analysis of these parameters confirms that the experimental data for the relatively long labels I, IV, and VI are in qualitative agreement with the theory of anisotropic rotation (see Chapter 1). According to this model, the properties of water–protein matrices are reflected in the parameters with the plus sign.

Table 10 lists the values of the empirical parameters $E^\pm$ and $\Delta S^\pm$ calculated from plots of log v as a function of $1/T$.

According to the values of the empirical parameters v, E^+, and k for Hb-β-93 SH-R at pH 7.1, we may conclude that the "leg" of the short spin label IV is held tightly in a fairly rigid matrix, whose structure varies only slightly with increasing temperature.[1] Similar values of these parameters are also observed for the short label in Hb-β-93 SH-R$_{VIII}$ (see rigid systems in Table 10). This conclusion is confirmed by the linear dependence of τ_c on the viscosity in the case of label VIII on hemoglobin in accordance with Stoke's law [64]. The values of the frequency of rotational diffusion for label VI bound to the β-93 SH group of hemoglobin (v), as well as the parameters E^+, ΔS^+, and k increase with increasing deviations of the pH from neutrality. According to the data obtained from the model systems

---

[1] The value of k is also slightly dependent on the temperature ($E = 3$ kcal/mole and $\Delta S^{\neq}_{eff} = -12$ e.u.).

Table 10. The Experimental Parameters $E^{\mp}$ and $S^{\mp}$ for Rotational Diffusion of Spin Labels Added to Proteins [81–83]

Protein, Label	pH	E^+ kcal	ΔS^+, e.u.	E^-, kcal	ΔS^-, e.u.
Hemoglobin β-93 SH	7.0	14.0	29	2.0	−12
I	—	(33)	(93)	—	—
	7.0	5.0	−8.9	—	—
IV	5.0	8.9	9.7	4.1	−8.8
VI	6.0	6.2	−0.1	3.1	−8.4
	7.1	4.8	−5.1	2.0	−12.2
	8.2	4.6	−5.3	5.0	−11.5
	—	(18.1)	(38.5)	—	—
	9.0	7.9	5.7	5.7	1.1
	—	(21.0)	(48.0)	(13.3)	(25.0)
	10.0	8.3	7.4	8.4	10.6
	—	(27.0)	(65.0)	—	—
	11.0	10.0	15.0	9.0	15.0
	—	(32.8)	(88.0)	(25.4)	(68.0)
Hemoglobin his (?), VII	7.0	5.8	−5.8	—	—
Myoglobin his					
VII	—	7.2	2.3	3.5	−7.0
	—	(11.0)	(26.0)	(6.4)	(2.4)
VIII	7.0	12.3	26.0	3.8	−4.1
	—	(36.8)	(103.0)	(4.8)	(−1.0)
XXVII	—	8.5	6.0	4.5	−2.0
	—	(11.5)	(18.0)	(6.4)	(4.1)
Myoglobin his (external),					
IV	—	9.7	12.4	4.0	−4.3
Myoglobin his (internal),	—	(17.0)	(39.0)	(5.1)	(−0.8)
IV	—	7.6	7.2	5.4	1.2
Lysozyme his	—	(10.7)	(17.5)	(3.4)	(4.3)
IV	5.5	17.0	40.0	6.2	3.0
	7.0	(18.6)	(46.0)	(6.2)	(2.5)
	9.0	13.5	27.0	5.2	−1.0
	11.0	(14.0)	(29.0)	(5.8)	(0.7)
	11.0	10.0	12.5	5.5	3.3
	—	(14.8)	(28.0)	—	—
	7.0	10.4	13.5	5.4	2.9
	—	(20.0)	(45.0)	—	—
VI	—	18.6	44.8	5.8	3.2
	—	(9.7)	(11.1)	(5.5)	(0.5)
VII	—	10.6	14.7	—	—
	—	6.3	−2.3	—	—
	—	(14.8)	(28.7)	(5.3)	(−0.7)
VIII	—	11.5	25.2	—	—
	—	(24.6)	(63.3)	(5.8)	(4.1)
Lysozyme lys, his, VII	—	5.6	−2.3	—	—
	—	(26.8)	(76.0)	(5.5)	(3.2)
Lysozyme asp-52, XXXI	—	7.7	13.8	—	—
	—	(11.7)	(21.2)	(4.9)	(1.0)

Note. The numbers in parentheses correspond to the high-temperature parameters beyond the discontinuities on the Arrhenius plots. The average in determining the parameters is 5–8%.

(see p. 128), this indicates the loosening of the protein matrix in acid and alkaline media. This loosening seems to involve some separating of the protein subunits, since the availability of the nitroxide tip of the label for encounters with paramagnetic probes increases at the same time (an increase in k).

The increase in k is not due to electrostatic effects, since the ratio k_w/k_p is practically independent of the charge of the paramagnetic probe (when potassium ferricyanide is replaced by dibenzenechromium iodide). The pH dependences of the measurable parameters ν, E^+, ΔS^+, and k, which reflect the changes in the conformational state of hemoglobin in the vicinity of the β-93 SH group, correlate with the pH dependences of the thermodynamic parameters for the binding of thiocyanate to the heme group. The experimental data on the thermodynamics of the binding of thiocyanate to hemoglobin and the activation parameters for rotational diffusion fall on one straight line [293] (see Fig. 53). At certain temperatures the Arrhenius curves for the spin labels show discontinuities, which indicate softening of the matrix. The transition temperatures range from 30 to 50° and are pH dependent. These discontinuities in the curves apparently correspond to the so-called "predenaturational transitions" that have been detected by various physical methods.

The increase in the length of the "leg" of the spin label in the transition from Hb-β-93 SH-R$_{VIII}$ and Hb-β-93 SH-R$_{IV}$ to Hb-β-93 SH-R$_I$ is accompanied by an increase in the values of ν, E^+, ΔS^+, and k characteristic of more pliable and softer systems (see p. 128), precisely as in the case of the loosening of the protein at acid and alkaline pH. In this case the label appears to interact with the surface layers of the protein and the structured water, both of which have high mobilities. The empirical parameters (Table 9) indicate that the shorter spin labels are bound to the rigid parts of the protein, while the longer labels are bound to the softer, conformationally more mobile parts of the water–protein matrix.

Lysozyme [83–113]. Egg lysozyme has been modified by spin labels of varying structure in different sections of the macromolecule (at the his-15 group by labels IV, VI, VII, and VIII and at the lys-13, lys-96, and lys-97 groups by label VII) (Fig. 48 and Table 1). According to the high values of k, the tips of the labels are readily available for collisions with probes; therefore, most of the time they are in an aqueous phase with a local viscosity close to the viscosity of pure water. This is also indicated by the fact that the EPR spectra at 77°K of spin-labeled lysozyme preparations and aqueous solutions of the radicals are identical. This would have been impossible if the nitroxide tips had been completely embedded in the hydrophobic matrix, which, as shown in the case of model systems, alters the EPR spectra of the radicals from those recorded in polar solutions.

In the case of lysozyme, as in the case of hemoglobin, there are significant differences in the values of $E^{\pm}$ and ΔS^{+}, determined from different components of the hyperfine structure. Here the values of E^{-} are considerably less sensitive to changes in the structure of the protein and the labels than is E^{+} (see Table 10). Thus the motion of the nitroxide group can be effectively described with the aid of a model of an anisotropic gyroscope that places the "leg" of the labels deep inside the protein matrix and the tip in the aqueous surface layer. The slower motion of the "leg" reflects the state of the matrix (the parameter E^{+}) to a greater extent than does the

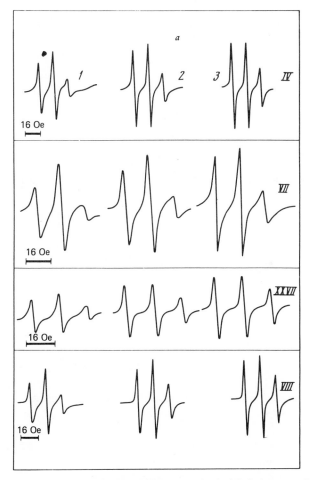

Figure 48. Electron paramagnetic resonance spectra of spin-labeled preparations of sperm whale myoglobin (a) and egg lysozyme (b) at various temperatures [32,44]. (a): (1) 16°; (2) 35°; (3) 55°; (b) (1) 15°; (2) 35°; (3) 55°; for the notation of the radicals, see p. 24.

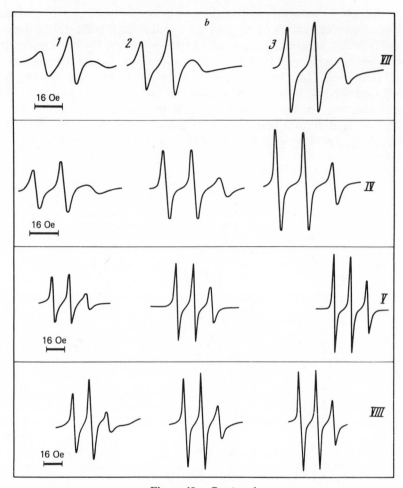

Figure 48 *Continued*

more rapid motion of the tip around the axis joining the piperidine ring to the "leg" (the parameter E^-).

The temperature dependences describing the rotation of labels on different lysozyme groups differ substantially from those for the free unbound labels in water (Fig. 49). The slope of the curves in Arrhenius coordinates and the nature of the discontinuities on the curves vary as functions of the structure of the label and the state of the protein (see Table 10). Several types of discontinuity can be noted on the curves. The first is detected at a temperature of 10° in the case of lysozyme 3 from the appearance of broadened spectral components in the low-temperature region. A second is observed at 20° in the case of lysozyme 2 and lysozyme 5 with spin-labeled ε-NH$_2$ lysine groups. Presumably, these are local structural changes in the vicinity of the lysine groups. The discontinuity on the curves at 50°, which is well defined in the case of lysozyme 4, is apparently due to the appearance of a more open conformation, which causes a decrease in, but not a complete loss of the enzymatic activity of, lysozyme in the high-temperature region [83,113]. All of the temperature-induced transitions are reversible.

In all cases the high-temperature region beginning from the discontinuity on the curves has higher values for E^+. The experimental parameters (small values of E^+ and ΔS^+) characterizing the motion of the nitroxide fragment of a short thick label on the his-15 group fit the comparatively rigid model systems such as cross-linked polymers with dense three-dimensional lattices. An increase in the flexibility of the label (lysozymes 1b and 1c) is accompanied by an increase in E^+ and ΔS^+ characteristic of softer, more pliable systems like glycerol and soft rubbers (see p. 128). Still higher values of E^+ and ΔS^+ are characteristic of long, flexible labels on lysozyme.

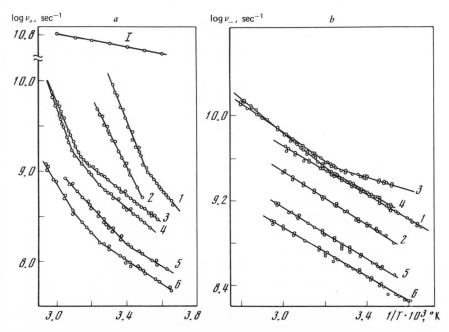

Figure 49. Arrhenius plots of the rotational diffusion parameters ν_+ (a) and ν_- (b) of spin labels on egg lysozyme at pH 7 [113]: (I) radical IV in water: (1) lysozyme 3; (2) lysozyme 1c; (3) lysozyme 4 (after freezing); (4) lysozyme 4; (5) lysozyme 2; (6) lysozyme 1a (for an explanation of the notation, see Table 1).

Careful consideration of the three-dimensional Phillips model for lyso-zyme leads to the conclusion that for almost all labels investigated there are practically no serious steric hindrances to rotation on the part of the protein groups. This is especially true of the spin labels on the relatively free, long flexible lys-96 and lys-97 residues. However, the experimental data indicate significant slowing of the nitroxide fragments, similar to the slowing of the radicals in glycerol solutions. This effect is apparently due to the special glycerol-like state of water on the protein surface. This layer forms as a result of the interaction of the side-chain groups and the surface polypeptide chains with the water molecules. Although the thermodynamic parameters of the glycerol-like water at each individual point do not appear to differ very much from the parameters of pure water, the total thermo-dynamic contribution, summed over many points can be very significant. There is a linear dependence in E^+ versus ΔS^+ coordinates (the compen-sation effect; see Fig. 53) for all of the lysozyme samples investigated.

Myoglobin [32,83].   Temperature dependences of the EPR spectra have been determined for five spin-labeled myoglobin preparations. These in-cluded four preparations modified by spin labels IV and VI–VIII, which have different lengths and flexibilities, at the same histidine groups and one preparations modified by spin label IV at the "internal"[1] histidine groups. All of the curves constructed for the five spin-labeled myoglobin preparations have a distinct discontinuity at 28° (Fig. 50), the longer spin labels having higher E^+ and ΔS^+ values and more abrupt discontinuities on the curves. At higher temperatures (>28°) the myoglobin macromole-cule undergoes a conformational change that is apparently due to the con-version of one conformer into another (A → B).

The reversible course of the Arrhenius curve from high to low tempera-tures indicates that, although conformers A and B are in dynamic equi-librium, as the temperature is varied one conformer is converted into the other. Thus at high temperatures conformer B predominates, and at low temperatures conformer A is dominant. The curve describing the depend-ence of the rotational parameters of the EPR spectra on the temperature in the case of spin-labeled myoglobin corresponds quantitatively to the curve describing the variation in the specific heat with increasing tempera-ture obtained by Privalov and Atanasov at the Institute for Proteins of the USSR Academy of Sciences (see Fig. 45) [32]. These curves are similar to the temperature dependence of the rate of magnetic relaxation of water

---

[1] The term "internal label position" is conventional, since there is no proof that the ni-troxide tips in the regenerated myoglobin are embedded in the protein globule. Neverthe-less, the procedure for adding the labels and the decreased rotation frequency indicate that internal labels interact with the more internal protein layers [32].

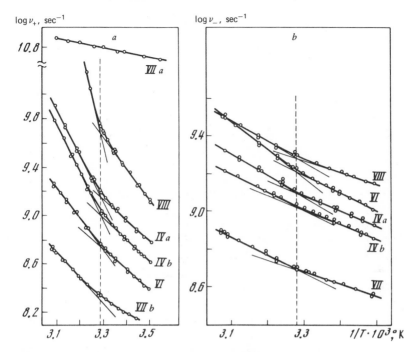

Figure 50. Arrhenius plots of the rotational diffusion parameters ν_+ (*a*) and ν_- (*b*) for spin labels on sperm whale myoglobin (radicals: VIIa, in water; VIIb, on protein; IVa, external; IVb, internal) [32]. (The notation of the radicals is explained on p. 24.)

protons. Thus the EPR-monitored local conformational changes in the vicinity of spin labels appear to be part of the overall conformational change in the macromolecule, which is monitored according to the total change in the specific heat and the rate of magnetic relaxation of the water protons. The experiments showed that the values of $k_w/2k_p$ for spin labels on proteins are 1.5–2.5 times higher than for the radicals in pure water. The differences in ν_+, ν_-, and $\nu_{-/+}$, in E^+, E^-, and $E^\pm$, as well as in ΔS^+, ΔS^-, and $\Delta S^\pm$ are due, in the final analysis, to differences in the motion of the parts of the spin labels directly in contact with the protein matrix and the motion of the nitroxide tips. The parameters ν_+, E^+, and ΔS^+ react with much greater sensitivity to changes in the state of the matrix than do the parameters ν_-, E^-, and ΔS^-.

According to the values of the parameters ν_+, E^+, ΔS^+, Δ and k for myoglobin, the "legs" of the short labels are tightly packed into a comparatively rigid matrix, whose structure varies only slightly with increasing temperature. The long labels seem to interact with the more surface layers of the protein, which possess greater conformational mobility and are easily softened with increasing temperature.

Comparison of the experimental data (Table 10) with the model data (see p. 128 and Table 9) implies that, with respect to the effective mobility in the vicinity of the long labels, the surface of the myoglobin macromolecule is a moderately viscous medium like a rubber with few cross-links. The values of E^+ and ΔS^+ are considerably higher, and the discontinuities on the curves are more well defined for the longer spin labels than for the shorter labels; therefore, the conformational changes are more large-scale.

The foregoing implies that the shell of the globule undergoes more changes than does the interior. There is a linear relationship between the experimental values of E^+ and ΔS^+ for spin labels attached to different functional groups of myoglobin (the compensation effect; see Fig. 53).

A compensational dependence of the ΔH on the ΔS of denaturation has also been observed [32] on the basis of experimental data on the heat of denaturation for a cyanide complex of myoglobin, which were obtained by optical, microcalorimetric, and other methods at various pH values.

Consideration of the myoglobin model with spin labels added to the surface histidine groups (see Fig. 30) leads to the conclusion that the protein groups should not exert any appreciable steric hindrance either on the rotation of the nitroxide tip or the free rotations along the bonds joining the "leg" of the spin labels to the polypeptide residues of the macromolecule [32,81,83]. However, experiments have shown that the motion of the nitroxide tips of labels is slowed by a factor ranging from 50 to 500, in comparison to the motion of the free unbound spin labels in solution. This immobilization is difficult to explain on the basis of electrostatic barriers arising from interactions between the polar groups of the spin labels and the charges on the protein. According to the Kendrew model, there are no electrostatic charges rigidly bound to the protein in the immediate vicinity of the labeled histidine groups. This is also indirectly supported by the results from tapping spin labels with paramagnetic charged probes with different signs (see Table 9). In addition, the nature of the temperature dependence of the rotational diffusion of the spin labels indicates that the longer labels have larger activation energies than do the shorter. At the same time the removal of the spin labels from the charged protein surface should have reduced the energy barriers to rotation of an electrostatic nature. Hydrophobic adherence of the nitroxide group to the protein, which should have caused a decrease in the values of E^+ and ΔS^+ (see Table 9), is also unlikely.

Therefore, at the present time the most reasonable explanation assumes the existence of a special state of water near the surface of myoglobin. This water appears to be comparatively rigidly structured in the first layer (e.g., near the electrostatic charges) and is gradually softened, acquiring a glycerol-like nature at a certain distance. Apparently, this water squeezes parts of the spin-label "legs," partially preventing free rotation of the labels around the bonds.

According to the data obtained with the aid of the spin probe–spin label technique, the local viscosity of the water corresponds to the viscosity of pure water, even at a distance of three-fourths of a layer of water.

Serum albumin [44]. In the case of human serum albumin (HSA) labeled with probe XLIII or spin labels I and VII under the same ambient conditions, the correlation time (τ) increases along the series: label I $<$ label VII $<$ probe XLIII. The low frequency of rotational diffusion (ν) for probe XLIII seems to be due to the packing of this radical in the deeper layers of the globule as a result of the hydrophobic nature of the interaction. Making the spin labels longer (I and VII) causes an increase in the distance between the nitroxide tip of the label and the portion that is bound, a decrease in the steric hindrances produced by the protein globule, and an increase in the probability of rotations around the internal bonds. It may be expected that the motion of probe XLIII in the systems studied most accurately reflects the properties of the protein matrix, while the rotational diffusion parameters of label I are primarily related to the properties of the medium surrounding the protein. Figure 51 presents plots of the rotation frequencies of radicals I, VII, and XLIII as functions of the viscosity of the solvent. The largest variation in mobility is observed in the case of label I, while the rotational diffusion of probe XLIII and label VII are practically independent of the viscosity of the medium.

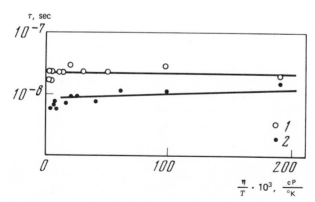

Figure 51. Variation in the rotational diffusion parameter (τ) of spin labels on proteins (10^{-3} M HSA) in sucrose as a function of η/T [44]: (*I*) label XLIII; (*2*) label VII.

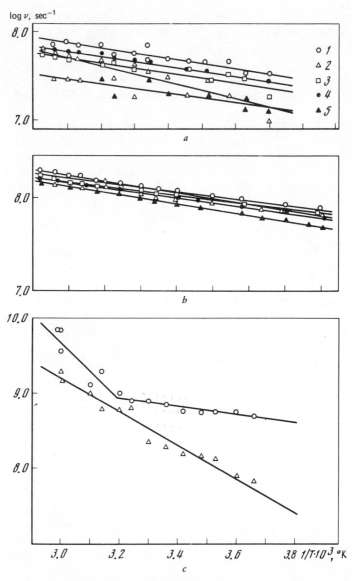

Figure 52. Arrhenius plots of the rotational diffusion parameters (ν) of spin labels and probes on 10^{-3} M HSA at various viscosities [44]: (*a*) spin probe XLIII; (*b*) spin label VII; (*c*) spin label I; sucrose addition (in %): (*1*) 0; (*2*) 30; (*3*) 40; (*4*) 50; (*5*) 60.

Table 11 presents the values of the effective energies (E_{eff}) and entropies ($\Delta S_{eff}^{\neq}$) for the diffusion of radicals I, VII, and XLIII. The E_{eff} values were calculated from the Arrhenius plots of log ν as a function of $1/T$ in Fig. 52.

Within the limits of experimental error, the activation energies of rotational diffusion for probe XLIII and label VII, unlike the E_{eff} for label I, are constants, being only slightly dependent on the viscosity of the surrounding medium and the degree of association of the protein molecules. Moreover the values of E_{eff} for probe XLIII and label VII coincide, despite the differences between them with respect to structure, the nature of the binding, and the absolute magnitude of ν.

Increasing the viscosity of the solvent by more than 50 times (60% sucrose) does not result in a change in the rotational diffusion of radicals VII and XLIII, while under these conditions the rotation frequency of the protein macromolecule as a whole decreases by a proportional number of times. A different pattern is observed in the case of spin label I, whose nitroxide tip is far from the protein matrix. According to the values of the experimental activation parameters for rotational diffusion (E_{eff} = 16.5 kcal and $\Delta S_{eff}^{\neq}$ = 39 e.u.), the motion of the radical in this case corresponds to motion in moderately viscous media (water–glycerol mixtures). At the same time, this motion is dependent on the bulk viscosity of the solution, and there are deviations from the Stokes–Einstein law. Similar dependences of τ_c on the viscosity were obtained for comparatively mobile labels on cytochrome and myoglobin [81,83].

Thus the mobility of radicals VII and XLIII, which are embedded within the protein globule, unlike the more freely bound radical I, is practically independent of the viscosity of the solvent and consequently of the motion of the protein globule as a whole. This conclusion is also confirmed by the results of experiments with suspensions in which the individual movements of the protein molecules and the radicals are hindered to an even greater extent, particularly as indicated by an analysis of the EPR spectra of the long radical I in the suspensions. The EPR spectra of these systems indicate strongly restricted rotational diffusion of the radical due to the high values of the viscosity within protein associates. Nevertheless in the case of albumin suspensions labeled with radicals VII or XLIII, which interact more directly with the matrix, there are no changes in the EPR spectra that could be attributed to a sharp increase in viscosity.

These results indicate that the motion of radicals embedded in globules is determined by the internal properties of the radical–protein system. The motion of radicals VII and XLIII may be treated either as the intrinsic rotation (twisting) of the radical relative to a solid protein matrix or as a consequence of the motion of sections of the protein matrix in the vicinity of the radical. The following arguments favor the second hypothesis:

1. The activation parameters for rotational diffusion of different radicals embedded in a protein globule are similar in value (see Table 11), that is, are independent of the nature of the radical and only dependent on the properties of the matrix.

2. The hydrophobic nature of the binding of probe XLIII assumes that there is contact between the probe and the protein globule at many points, which would make the twisting of the radical relative to the solid walls of the matrix difficult. Evaluation of the energy (E) of the interaction between probe XLIII and the walls of the matrix due to dispersion forces from the value of the enthalpy of sublimation yields the value $E > 10$ kcal, which is considerably higher than the experimental values $E_{eff} < 3$ kcal.

3. The expected value of the preexponential factor for rotational diffusion of a molecule in a rigid cell must be close to 10^{13} sec^{-1}. However, the low values of the preexponential factors and E_{eff} (see Table 11) indicate that the motion of the medium around the radical is complex. The qualitative model of this motion treated in the concluding part of this chapter attributes the decreased values of E_{eff} and $\Delta S_{eff}^{\neq}$ to the need for compression of the adjacent portions of the protein matrix in order to create a free volume sufficient for the movement of the probe itself or groups in close contact with it. Thus, explaining the laws governing the rotational diffusion of radicals VII and XLIII as a consequence of the local mobility of the protein matrix, which is primarily due to the segmental motion of the polypeptide chains, is very likely. Investigations of the motion of the albumin protein matrix with gamma-resonance labels (Chapter VIII) have demonstrated the displacement of Mössbauer nuclei attached to the polypeptide chains in the absence of motion in the HSA macromolecules and lead to the same conclusions.

Active centers of enzymes. The introduction of nitroxide radicals into the active centers of enzymes opens up considerable new possibilities in the study of the local mobility and the local viscosity of a medium in which acts of sorption and chemical conversions take place under enzyme action.

Even in the first paper by Berliner and McConnell [104] it was pointed out that spin label XII, which has a very short "leg," is greatly immobilized when placed in the active center of α-chymotrypsin due to the strong interaction with the walls of the protein matrix. These workers assumed that the combination of the ease in modifying the ser-195 group of the enzyme and the strong immobilization of the label can be explained by the moderate rigidity of the matrix, which allows small fluctuations in the size of the cleft of the active center to accommodate the radical.

Table 11.   Values of the Effective Energies (E_{eff}) and Entropies (ΔS_{eff}) for Rotational Diffusion of Radicals I, VII, and XLIII Bound to HSA at pH 7 and 20° [44]

Radical	Sucrose	E_{eff}, kcal/mole	$\Delta S_{eff}^{\neq}$, kcal/mole·deg
XLIII	0 (H_2O)	2.5±0.3	−15.4±0.5
XLIII	30	2.5±0.2	−14.6±0.5
XLIII	40	2.5±0.2	−15.4±0.5
XLIII	50	2.5±0.2	−15.4±0.5
XLIII	60	2.3±0.3	−15.9±0.2
VII	0 (H_2O)	2.4±0.5	−12.7±0.2
VII	30	2.1±0.1	−13.6±0.2
VII	40	1.8±0.1	−14.7±0.2
VII	50	2.0±0.1	−14.0±0.2
VII	60	1.9±0.1	−14.3±0.2
I	0 (H_2O)	16.5±1.5	39±5
I	60	10.2±0.5	15±2

Note.   The values of the rotation frequency (ν, sec^{-1}) for the radicals are: XLIII, 3×10^7; VII, 1.2×10^8; I, 5.9×10^8 (water) and 1.6×10^8 (sucrose).

McConnell subsequently studied the viscosity dependence of the rotational diffusion of spin label XII and the large radical XXI covalently bound to the serine group of α-chymotrypsin [65]. It turned out that at low viscosities (< 10 cP), the rotation rate of the labels is in satisfactory agreement with Stokes's law ($\tau_c = 11.5 - 13 \times 10^{-9}$ sec in water) and is close to the rotation rate of the α-chymotrypsin macromolecule. A linear dependence was observed in T/η versus τ_c^{-1} coordinates up to the moment at which $\tau_c = 4 \times 10^{-8}$ sec, indicating that the local motion of the sections of the protein interacting with labels XII and XXI, where such motion takes place, is characterized by a larger correlation time than is the rotation of the macromolecule.

A noncovalent specific interaction between the active center of α-chymotrypsin and a spin-labeled hydrophobic inhibitor was used to study the mobility of this active center [27]. After incubation of the probe and enzyme in a phosphate buffer at pH 3.0, two types of signal were observed with rotational correlation times of $\tau_1 = 17 \pm 5 \times 10^{-9}$ sec and $\tau_2 < 10^{-10}$ sec. The decrease in the intensity of the signal from the immobilized radical in the presence of acridine and cinnamoylimidazole, as well as the competitive nature of the slowing of the hydrolysis of the ethyl ether of L-tyrosine under the effect of radical XXXIII ($k_I = 2 \times 10^{-4}$ M), indicate the binding of radical XXXIII to the hydrophobic portion of the active center of α-chymotrypsin. The value of τ_c is only slightly dependent on the viscosity of the medium ($\leq 60\%$ sucrose), the temperature (15–40°), and the precipitation of the protein with ammonium sulfate in the presence of radical XXXIII. Thus the motion of the spin probe is only slightly dependent on the behavior of the macromolecule as a whole and is apparently regulated by the mobility of the hydrophobic portion of the active center relative to the protein globule with a correlation time approximately equal to 10^{-8} sec. Since the motion of spin probes generally reflects the motion of the local environment [2,38], and the probe is bound to the enzyme by many hydrophobic contacts, the possibility of intrinsic motion (twisting) of the probe relative to the rigid hydrophobic portion is unlikely. The values of the effective energies and entropies of activation for the motion of the probe equal, respectively, 2 kcal and -18 e.u.

A different pattern is observed for the rotational diffusion of spin labels on the met-192 group [288] (radical IV) and the ser-195 group [107] (radical XV) of α-chymotrypsin. Spin labels IV and XV are joined to the enzyme by flexible "legs" that allow rotation of the nitroxide tips relative to the protein matrix. This rotation is characterized by a much greater temperature coefficient than the motion of the protein macromolecule (Table 12), and the values of the parameters E_{eff} and $\Delta S^{\neq}_{\text{eff}}$ correspond to the motion in the glycerol-like water–protein layer.

The same effects have been revealed qualitatively for spin-labeled ribonuclease A [110]. The values of the effective activation energy for rotational diffusion of spin label XXII, which is positively located on the his-119 group of the active center and the calculated values of $\Delta S^{\neq}_{\text{eff}}$ in the high-

Table 12.   Values of the Effective Activation Parameters for Rotational Diffusion of Spin Labels and Probes Introduced into the Active Centers of Enzymes and Oxygen Carriers

Enzyme, label	E_{eff}, kcal	ΔS_{eff}, e.u.	Reference
α-Chymotrypsin			
ser-195, XV	8.7	11	[107]
met-192, IV	11.3	20	[288]
hydrophobic pocket, XXXIII	2.0	-18	[27]
Ribonuclease A, his-119, XXII			
$T > 35°$	10.8	19	[110]
$T < 35°$	5.5	4	
Myoglobin, heme, XLVIII	2.8	-14	[Unpublished data by G.I. Likhtenshtein, L. W. Ivanov, and T. V. Avilova.]

temperature region, fall within the range of values characteristic of glycerol-like systems.

The local mobility of the protein matrix in the vicinity of the heme group of myoglobin was investigated with the aid of the isocyanide spin label XLVIII. The label was introduced into a single crystal by prolonged incubation and thereby immobilized [288]. The Kendrew model clearly shows (see Fig. 30) that the nitroxide tip of the label can penetrate into the active center only if the polypeptide chains making up the walls of the cleft of the active center are considerably separated (by 3–4 Å). It is clear that the nitroxide group in the single crystal should be strongly held by the protein matrix and that the motion of the macromolecule is stopped. Nevertheless the EPR spectra of the spin-labeled myoglobin single crystal show that the nitroxide fragment of the label possesses some mobility ($\nu = 5 \times 10^7$ sec^{-1}, $E_{\text{eff}} = 2.8$ kcal, and $\Delta S_{\text{eff}}^{\neq} = -14$ e.u.).

This mobility is difficult to explain without assuming the existence of local thermal motions of the protein matrix, which result from time to time in the drawing apart of the walls of the cleft and in rotation of the nitroxide fragment into another position relative in the protein macromolecule.

Thus the quantitative study of the rotational diffusion of spin labels introduced into the active centers of enzymes has revealed two qualitatively different cases. In one case tightly held labels have limited mobility, which seems to reflect the local mobility of the protein matrix. In the other case the nitroxide tips of longer labels rotate in a moderately viscous water–protein medium. The experimental values of E_{eff} and $\Delta S_{\text{eff}}^{\neq}$ (Table 10 and Fig. 53) for the active centers of different enzymes fall on a straight line in E_{eff} versus $\Delta S_{\text{eff}}^{\neq}$ coordinates and thus demonstrate a compensation effect similar to that observed in enzymatic reactions (Chapter 9). This fact is an argument supporting the fact that the compensation effect in enzymatic reactions is caused by changes in the structure of the water–protein matrix of the enzyme during the elementary steps in the chemical reactions.

Some general remarks concerning the method for determining local mobility in water–protein matrices. The experimental data cited in this chapter indicate that the nitroxide fragments of labels and probes on proteins generally undergo complex motion. Since the EPR spectra of nitroxide radicals primarily show the most rapid correlation mechanisms, there is the possibility of at least a semiqualitative analysis of the individual types of label motion. One such possibility involves varying the overall viscosity of the medium and the aggregation state of the protein macromolecules. For comparatively long labels, which move rapidly relative to the matrix,

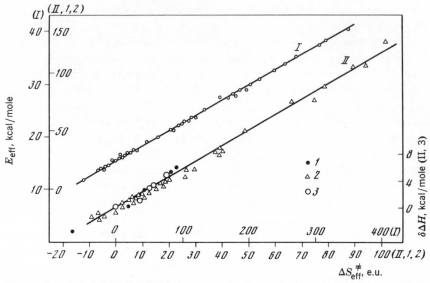

Figure 53. Variation in the activation energy (E_{eff}) as a function of the entropy of activation ($\Delta S_{eff}^{\neq}$) [193]: (*I*) denaturation of proteins; (*II*) rotational diffusion of spin labels on proteins and enzymes: spin labels in the active centers of enzymes (*I*); data from Tables 10 and 12 (*2*); relative changes in the thermodynamic parameters ($\delta \Delta H$) for the binding of isothiocyanate by the heme groups of hemoglobin (*3*).

an increase in the rotational correlation time τ_c should be expected with increasing overall viscosity, in qualitative agreement with Stokes's law.

In the case of radicals tightly held in a protein matrix, Stokes's law should be fulfilled quantitatively over a broad range of viscosities, and the effective radius of rotation of the label must coincide with the radius of the globular macromolecule. If a tightly bound nitroxide radical moves more slowly than the macromolecule, as the viscosity is increased, we should first expect the rotation of the macromolecule to obey Stokes's law. When a certain critical viscosity value is reached, the value of τ_c should no longer be dependent on the properties of the medium or the aggregation state of the protein. These cases have in fact been observed experimentally.

The laws governing the activation parameters for the diffusion of spin labels on proteins can be explained within the framework of the following simplified model. Let us assume that an act of rotation of a nitroxide fragment can occur only after the rate-limiting step for rearranging the surrounding portion of the water–protein matrix, which in the simplest case consists of n almost identical unit cells (water molecules or monomers). Then in the case of a reversible rearrangement, we can write the following equations for the effective energies and entropies of activation [289,293]:

$$E_{\text{eff}} = E_i + n \cdot \Delta H_0 \tag{7.2}$$

$$\Delta S_{\text{eff}}^{\neq} = \Delta S_i^{\neq} + n \cdot \Delta S_0 + 2.3 \log a, \tag{7.3}$$

where ΔH_0 and ΔS_0 are the enthalpy and entropy for a rearrangement in one member of the ensemble, E_i and $\Delta S_i^{\neq}$ are, respectively, the energy and entropy of activation for the first step in initiating the rearrangement, and the value of a reflects the way the monomers are packed in the ensemble [289].

If the rearrangement involves unordering ("melting") of the surrounding matrix, then ΔH_0 and ΔS_0 are greater than zero, and overestimated values of $\Delta S_{\text{eff}}^{\neq}$ and E_{eff} are observed experimentally. This seems to be the case for comparatively long labels whose motion is dependent on the reorientation of a loose water–protein matrix. In the comparatively densely packed deeper layers of the protein matrix, the creation of a free volume for the rotation or local movement of the section of protein with a label occurs at the expense of additional compression ("crystallization") of the medium. This case yields negative values for ΔH_0 and ΔS_0 and thus underestimated values of E_{eff} and negative values of $\Delta S_{\text{eff}}^{\neq}$. Such parameters are in fact characteristic of radicals implanted in a protein matrix. According to Eqs. (7.2) and (7.3), a linear relationship (compensation effect) should be observed for systems differing with respect to the size of the cooperatively rearranged segment (n), and we thus have

$$E_{\text{eff}} = \alpha + \beta \cdot \Delta S_{\text{eff}}^{\neq}, \tag{7.4}$$

where $\alpha = E_i - T_0 \cdot \Delta S_i^{\neq}$ and $\beta = (\Delta H_0 / \Delta S_0) = T_0$. Here T_0 is the "melting" or "crystallization" point of the surrounding medium.

In fact, the values of E_{eff} and $\Delta S_{\text{eff}}^{\neq}$ for the spin-labeled proteins studied fall on one straight line, demonstrating a universal compensation effect (see Fig. 53). The parameters of the compensation effect for the rotational diffusion of spin labels are practically independent of the size of the protein macromolecules and the structure of the labels. This again confirms the hypothesis that the diffusion parameters effectively reflect the local state of the water–protein matrix in the vicinity of bound radicals.

We can, thus, distinguish the following three surface layers in protein matrices (Fig. 54): (a) the outer aqueous layer with properties similar to the properties of pure water, (b) the water–protein matrix, the motion that corresponds to the motion in glycerol or soft rubbers and that contains the side-chain groups of proteins and water molecules in clefts, and (c) the deeper layer, which includes the outer polypeptide chains, side-chain groups that are squeezed by the matrix, and strongly bound water.

The quantitative laws governing the activation parameters for rotational diffusion of spin labels on proteins are similar to the laws observed in the

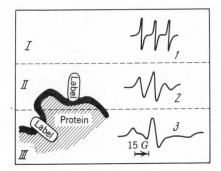

Figure 54. Mobility of various layers of a water–protein matrix: (*1–3*) typical EPR spectra of nitroxide radicals submerged in the respective matrix layers; (*I*) outer aqueous layer ($\nu \sim 10^{11} - 10^{10}$); (*II*) glycerol-like matrix layer ($\nu \sim 10^{10} - 10^8$); (*III*) relatively solid matrix layer ($\nu \sim 10^8 - 10^7$) [193]; ν is the rotational diffusion correlation frequency of the radicals (in sec^{-1}).

denaturation of proteins as well as in certain enzymatic reactions. All the systems mentioned above show a similarity in the linear enthalpy–entropy correlations (the compensation effect), which is readily explained in the context of models that assume rapid reversible rearrangements of the water–protein matrix accompanying the elementary chemical steps in enzymatic reactions and acts of rotational diffusion in spin labels. The possible involvement of the local mobility of the various layers of protein matrices in bringing about the catalytic properties of enzymes and its effect on the thermodynamic and kinetic parameters of enzymatic reactions is discussed in Chapter 9.

Chapter Eight
Luminescent, Mössbauer, and Electron-scattering Labels

In recent years other techniques based on the labeling concept have been used along with spin labeling to investigate biological structures, primarily enzymes. The experience in the combined use of luminescent, Mössbauer, and electron-scattering labels along with spin labels gained at the Laboratory for the Kinetics of Enzyme Action at the Institute of Chemical Physics of the USSR Academy of Sciences may be useful for investigating biological systems of various degrees of complexity [12–18].

A complete presentation of all the theoretical and practical aspects of these methods is beyond the scope of the problems discussed in this monograph. In this chapter we shall present only the basic principles and results. The techniques first proposed or most completely developed at the Laboratory for the Kinetics of Enzyme Action will be described in somewhat greater detail.

LUMINESCENT LABELS

The introduction of luminescent chromophore groups into proteins has long been in practice in biochemical and biophysical investigations. The basic parameters of luminescent chromophores, namely, the luminescence intensity (I), the lifetime of the excited state (τ), the degree of depolarization (p), the wavelength for the fluorescence maximum $(\lambda_{max})m$, and the shape of the spectra, are greatly dependent on such factors as the structure of the local environment, the rotation rate of the macromolecules, and the presence of other chromophores. Various aspects of the application of luminescence techniques in biological investigations have been presented in detail in several monographs and reviews [295–302]. Here we shall dwell only on the technique for investigating structure by means of a quantitative measurement of the migration of the energy of an excited state. In our opinion, the extensive possibilities of this technique have been underestimated by biochemists.

The determination of distances between functional groups in biological structures can be based on the transfer of energy between chromophores by an inductive resonance mechanism, since the probability of energy transfer is dependent on the distance (r) between the chromophores.

Energy transfer is established experimentally by the following methods: (a) according to the quenching of the luminescence of the donor groups (D) in the presence of acceptor groups (A), (b) according to the depolarization of the luminescent groups due to the fact that, following the transfer of a quantum of polarized light from one molecule (D) to another (A), there is a change in the polarization density of a secondarily scintillated quantum, and (c) according to the decrease in the lifetime of an excited state D.

In principle, there are several possible variants of the method.

1. The luminescent donors are intrinsic protein groups (tryptophan and tyrosine), and the acceptor is a quenching label.
2. The level luminesces, and a chromophoric protein group (e.g., the active center) acts as the quencher.
3. The acceptor and donor functions are performed by labels. Migration between luminescent groups is a special case of this variant.

The phenomenon of inductive resonance energy migration was first discovered by Vavilov [299], and the quantitative theory was developed by Förster [300] and Galanin [301]. The relationship between the experimentally determined fluorescence intensities of the acceptor i in the presence (I) and absence (I_0) of the quencher j and r_{ij} is defined by the formula

$$\frac{I}{I_0} = \sum \frac{1}{1 + \sum \dfrac{R^6}{r^6{}_{ij}}}, \tag{8.1}$$

where R is the so-called "critical radius," which is equal to the distance at which the probability of energy migration from the excited donor level becomes equal to the probability of emission. According to quantum mechanics, the value of R can be determined with the aid of the following expression

$$R^6 = \frac{9 \cdot 10^3 \ln 10 K^2 \varphi}{128 \cdot \pi^5 n^4 N_0} \int_0^\infty \frac{f(\nu) \cdot \varepsilon(\nu)\, d\nu}{\nu^4}\ \mathrm{cm}^6, \tag{8.2}$$

where K^2 is an orientation factor, which equals $2/3$ for a random orientation; n is the index of refraction of the medium, which is protein in this case; $f(\nu)$ is the distribution function of the fluorescence intensity normalized by the quantity $\int f(\nu)\, d\nu$; $\varepsilon(\nu)$ is the distribution function in the absorption spectrum in terms of the extinction coefficient ($\mathrm{M^{-1} \cdot cm^{-1}}$); φ is the fluorescence quantum yield of the donor; and N_0 is Avogadro's number.

For an approximate calculation of R we can use the following formula, which takes into account the specific nature of proteins [302]:

$$R^6 = 0.66 \times 10^{-33} \tau_0 I / \nu_0^2\ \mathrm{cm}^6, \tag{8.3}$$

where τ_0 is the lifetime of the excited donor state, ν_0 is the average value of the wavenumbers of the absorption maxima of the donor and the acceptor, and I is the overlap integral for the fluorescence spectra of the donor and the absorption spectrum of the acceptor, whose intensities are expressed in terms of the extinction coefficient. In the case of one donor, Eqs. (8.1)–(8.3) can be used to calculate the distance to the acceptor r, and for a system of n donors we have

$$(\bar{r})^{-1} = \sqrt[6]{\sum_i \frac{1}{r^6_i}}, \qquad (8.4)$$

where r_i is the distance from the acceptor to the ith donor. The values of τ_0, φ, and n are usually known for the most widely employed luminescent chromophores; in other cases they can be found experimentally. The quantities $\bar{\nu}_0$, $f(\nu)$, $\varepsilon(\nu)$, and I are determined from the experimental absorption spectrum of the acceptor and luminescence spectrum of the donor. In the general case the effect of luminescence quenching in a solution can be due to various causes: (a) a chemical interaction between the fluorescing and quenching groups, (b) shielding, (c) reabsorption, (d) active collisions between the donor and the acceptor, (e) a triplet–triplet transfer of energy from the donor to the acceptor, and (f) a single–singlet transfer of energy by an inductive resonance mechanism.

However, carrying out a whole series of experiments and introducing the necessary corrections makes it possible to isolate the contribution to the quenching from the transfer of energy by an inductive resonance mechanism. Energy transfer by this mechanism is usually predominant in dilute solutions of proteins and when the donor and acceptor centers are separated by distances greater than 10 Å. The details of the calculation can be found in various works [11–13,93,94,303,304]. In practical applications of Eqs. (8.1–(8.4) we cannot avoid dealing with the indefinite nature of the value of K^2, since the relative orientation of the chromophores is generally unknown. However, since R depends on the cube root of K, the deviation of K^2 from $2/3$ has a significant effect only for extreme (perpendicular and parallel) orientations. In real systems there is a set of orientations due to the flexibility of the "legs" of the labels and the existence of several vectors for electronic transitions in complex metals. Therefore, calculations with the aid of Eqs. (8.2) and (8.3) with $K^2 = 2/3$ seem to yield fairly accurate results. This is indicated, in particular, by the data on the study of energy migration in spin-labeled hemoglobin [13].

Luminescence-quenching labels. This method was developed with the aid of bovine (BSA) and human (HSA) serum albumin, which possess tryptophan and tyrosine fluorescent centers. The chromophore residues of the

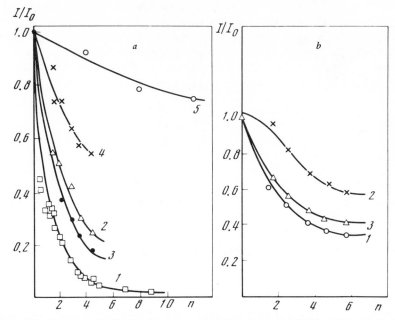

Figure 55. Variation in the luminescence-quenching intensity (I/I_0) of 1.2×10^{-5} M BSA as a function of the number (n) of molecules of the dye 4RS bound at pH 6.35 and 20° to the protein molecule [12]. Fluorescence at $\lambda = 340$ nm (tryptophan) (a) and $\lambda = 300$ nm (tyrosine) (b). Measurement conditions: (*1*) pH 6.3; (*2*) incubation at 100° for 5 min; (*3*) pH 9.8; (*4*) treatment with 8 M urea at pH 6.3; (*5*) addition of the hydrolyzed dye to BSA denaturated with 8 M urea at pH 9.8.

trichlorotriazine dyes 5BC, 2RP, and 4R1 served as the acceptor groups [12].

The quenching intensity (I/I_0) is strongly dependent on the number of labels bound and on factors that disturb the conformation of the proteins (Fig. 55). The results of the calculation of the distance r between a single tryptophan group of human serum albumin and the quencher molecules with the aid of Eqs. (8.1)–(8.4) are presented in Fig. 56. As expected, the most substantial increases in r (5–12 Å), which reflect conformational changes causing loosening of the protein macromolecule, are observed under the action of 8 M urea. More modest changes in r (2–4 Å) occur when the protein is deformed as the result of acid–base transitions in the protein groups. Similar phenomena are observed in studying the migration of energy between the tyrosine residues of albumin and the labels.

There is some special interest in quencher labels that contain paramagnetic ions (e.g., the copper-containing procion dye 2RP). The applica-

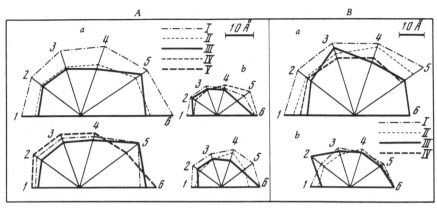

Figure 56. Diagrams of the distances (r) and average distances ($\bar{r}$) between the tryptophan (a) and tyrosine (b) residues of HSA (A) and BSA (B) and between molecules of the dye 4RS added to the protein [12]: (1–6) number of dye molecules added; angles selected arbitrarily; (A) (r): (I) treatment with 8 M urea; (II) thermal treatment (60°, 1 hr); (III) pH 6.3; (IV) pH 3.4; (V) pH 9.8; (B) ($\bar{r}$): (I) treatment with 8 M urea at pH 6.3; (II) thermal treatment (100°, 5 min); (III) pH 6.3; (IV) pH 9.8.

tion of these compounds permits simultaneous evaluation of the distances between the luminescent center and the label from the luminescence quenching and between the labels from the dipolar broadening of the EPR spectrum.

Quenching the luminescence of labels. This method was first employed by Weber and Teal [11] to study the luminescence quenching of hemoglobin modified at its lysine and histidine groups by a fluorescent dye. The great number of these groups allowed these workers to use a statistical approach.

The possible uses of this method to evaluate distances between labels and active centers of enzymes in the case of rigorously fixed labels have been analyzed experimentally [13]. This variant makes it possible to determine distances between groups in active centers and to study allosteric conformational changes. A convenient object for adding luminescent labels is the β-93 SH group of bovine hemoglobin, whose exact position relative to the quenching heme group is known from the Perutz X-ray diffraction model. Experience has shown that the addition of an isoimidorhodamine label to hemoglobin is accompanied by intense fluorescence quenching, which, according to a calculation employing Eqs. (8.1)–(8.3), corresponds to 19 Å. Measurement of the distance between the center of the heme group and the chromophore center, averaged for the possible conformations with the aid of the Perutz model, also yields a value of about 20 Å.

The value of r has proved to be sensitive to various effects on the active center of hemoglobin and the state of the macromolecule. For example, carboxylation causes an appreciable decrease in r from 19 to 17 Å. Denaturation with urea increases r to a value greater than 26 Å. When the pH is shifted from neutral values, there is a sharp decrease in the degree of quenching, apparently due to the separating of sections of the protein.

A similar approach was recently used in the work of Alfimova, Syrtsova, and the present author to evaluate the distance between the SH group in the ATPase center of nitrogenase and the nonheme iron (Fig. 57) [304].

The distribution of eight iron atoms over the macromolecule of a model nonheme iron–sulfur protein was studied in [93]. According to Eqs. (8.1)–(8.4), when the distribution of the luminescent labels over the surface of a specific macromolecule is random, the value of $\bar{r}$ as well as that of I/I_0 must be dependent on the relative positions of the quenching centers, in this case the iron atoms. Thus, when the arrangement of the iron atoms in the protein globule is uniform or diffuse, the probability that there are luminescent labels in the quenching zone will be greater than when the atoms are crowded into a small part of the protein globule. Comparison of the experimental values of I/I_0 with the theoretically expected values calculated by Weber and Teal's method [11] shows that the experimental value of I/I_0 corresponds most accurately to the variants in which the iron atoms are in a group of eight atoms.

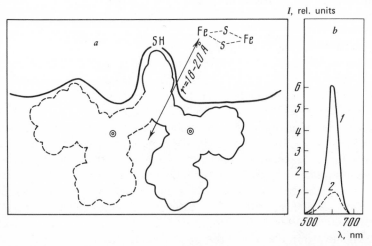

Figure 57. Structure of a fragment of the active center of nitrogenase (a) [304] and luminescence spectra of an isoimidorhodamine label on the SH groups of the ATPase center of nitrogenase (b) [304]: (1) label in solution; (2) after addition of the label to nitrogenase.

These examples illustrate the prospects of using luminescent and quenching labels to solve various problems regarding the structure of biological systems. The high sensitivity of luminescence quenching resulting from the sharp dependence of energy migration on the distance between centers suggests that the quenching method will become widely employed.

NUCLEAR γ RESONANCE (THE MÖSSBAUER EFFECT)

This method is based on the transition of a nucleus from the ground state to an excited state upon absorption of a specific amount of electromagnetic radiation (γ quanta). These transitions are similar to the corresponding transitions between electronic molecular levels that result in the absoprtion of light in the ultraviolet and visible regions. In either case the basic resonance condition $h\nu = \Delta E$ must be fulfilled, but the values of ΔE for the nuclear levels are very high (10^3–10^5 eV). Nuclear γ resonance can be observed only when the emitting and absorbing nuclei are bound fairly strongly to a matrix, and the energy given off is assimilated by the crystal lattice as a whole [309–321].

Experience has shown that relative displacements of the nuclear levels and the extent to which they are blurred reflect the chemical and physical structure of the environment with great sensitivity. The principal parameters measured experimentally are the chemical shift (δ), the quadrupole splitting (Δ), the absorption intensity, the lineshapes, and the hyperfine structure of the γ-resonance lines. The values of δ for compounds of trivalent iron (the d^5 state) range from -0.3 to $+0.3$ mm sec^{-1} for covalent complexes and from 0.1 to 0.6 for ionic complexes, while in the case of divalent iron (d^6) the values of δ range from 0.7 to 1.4 mm sec^{-1} [301], which corresponds to the additional shielding of the $4s$ electrons upon addition of an electron to the $3d$ orbital. Another important parameter of this phenomenon is the quadrupole splitting (Δ), which arises as a result of the interaction between the nuclear quadrupole moment and the nonuniform electric fields of the chemical environment. For example, in the case of complexes with axial symmetry, the gradients of the electric fields along and across the axis are different, causing the existence of two types of absorbing nuclei and the quadruple splitting of the γ-resonance lines. A definite Lorentzian lineshape is characteristic of individual complexes. If there are a number of complexes with similar parameters in a system, the spectrum consists of several superimposed lines.

In those cases in which a ^{57}Fe nucleus is in an external or internal magnetic field (i.e., if the Mössbauer atom is paramagnetic), the γ-resonance spectrum may show a magnetic hyperfine structure. In the ground state the ^{57}Fe nucleus has a magnetic moment with $I = \frac{1}{2}$, which can be oriented either along a magnetic field or against it. The magnetic moment of the excited ^{57}Fe nucleus ($I = \frac{3}{2}$) can assume the following four orientations: (a) along the field, (b) against the field, and (c) at different angles to the field. Differently oriented magnetic moments will make different contributions to the interaction energy; therefore, the γ-resonance spectrum consists of six or more components. The distance between adjacent components is proportional to the magnetic field strength H, permitting calculation of its value in the nucleus. It turns out that in ferromagnetic and antiferromagnetic compounds with strong interactions between the unpaired electrons, the strength of the intrinsic magnetic field H reaches enormous values, even up to 600,000 Oe. Other paramagnetic centers that influence the local fields of an unpaired electron may be detected from the degree to which they affect the hyperfine structure of the spectra.

Evaluation of distances between centers by nuclear γ resonance. The γ-resonance spectra of Mössbauer atoms (e.g., ^{57}Fe) are very sensitive to the electronic structure and symmetry, and not only of the immediate environment. Under certain conditions, γ-resonance spectra can reflect more distant interactions, such as spin-exchange and dipolar effects.

In various studies [81,93,94,317] this fact was exploited to establish the multinuclear nature of nonheme iron–protein complexes with ^{67}Fe atoms introduced into their active centers. It is known [309–317] that the γ-resonance spectra of individual multinuclear paramagnetic complexes show an increase in the relaxation time of the spins with decreasing temperature, and a magnetic hyperfine structure (e.g., six lines instead of two) appears as a consequence. This is due to the interaction between the spins of the electron and the nucleus, which is averaged at higher temperatures due to the rapid relaxation of the spin. However, in the case in which the iron ions are separated by short distances, a magnetic hyperfine structure cannot appear as a result of spin–spin interactions (the pairing of spins or a decrease in the electronic relaxation time). This phenomenon may be used to study the relative positions of iron ions by measuring the temperature dependence of the γ-resonance spectra.

Experiments carried out at the temperature of liquid helium have shown that the γ-resonance spectra of the ^{57}Fe nuclei in a model nonheme protein, paramagnetic nitrosyl complexes of it, and the enzyme nitrogenase show no magnetic hyperfine structure, apparently because of the close arrangement of the iron ions in these biological structures. A similar approach was used to established the binuclear nature of the active center of ferredoxin. The close arrangement of the iron atoms in the protein has been confirmed in all cases by independent methods, primarily with spin labels (Chapter 3).

Investigation of the dynamic structure of proteins with γ-resonance labels. Spin labels have been widely employed to monitor conformational changes in proteins (Chapters 6 and 7). However, that method permits reliable detection of the motion of labels only with frequencies ranging from 10^7 to 10^{10} sec^{-1}. In addition, the large dimensions of the spin labels make it impossible to directly follow the molecular motions of protein groups.

Considerable information on the dynamics of the local movements in individual portions of a globule can be obtained with the aid of nuclear γ-resonance when compounds containing Mössbauer isotopes serve as labels. The undirected motion of the label (which is similar to diffusion) must produce broadening of the resonance-absorption line, if the correlation time of the motion is equal to or greater than the lifetime of the nucleus in the excited state. In addition, an increase in the vibrational amplitude of a Mössbauer nucleus in the direction of the propagation of the γ-quantum results in a decrease in the probability of nonradiative absorption (f') [309]. Thus the mobility of Mössbauer labels added to a specific group must have a significant effect on the spectrum of the γ-resonance labels (Fig. 58).

The parameters of γ-resonance spectra, namely, the intensity of the lines (f') and the relative line broadening ($\Delta G/G$) (see Fig. 58), must be dependent on the parameters of the local motion of the Mössbauer atom. These are x^2, the square of the amplitude of the displacement of a nucleus upon emission (in Å), and D, the diffusion coefficient of a particle rigidly bound to a ^{57}Fe ion, which is primarily dependent on the temperature, the viscosity in poise, and the effective rotation radius (a) in angström units [309–324]. Here we have

$$f' \simeq \exp(-53x^2) \qquad (8.5)$$

and

$$\frac{\Delta G}{G} \cong \beta D = 100 \frac{T^0 \text{K}}{\eta a} \qquad (8.6)$$

For example, displacement of the nucleus by 1 Å causes a drop in the probability of the effect by more than a factor of 10,000. The γ-resonance line of a Mössbauer nucleus rigidly bound to a macromolecule with a radius of 30 Å in an aqueous solution must be broadened by almost 10^6 times. Thus media with high viscosities and small solid matrices are necessary to observe γ-resonance spectra. It is clear that the local displacement of a proton matrix can be monitored by nuclear γ-resonance only if the protein macromolecules are rigidly fixed in the crystalline state or in a frozen solution.

The possibilities of applying γ-resonance labels to the study of the dynamic structure of proteins have been analyzed in the case of human serum albumin modified with ^{57}Fe ions [15,16].

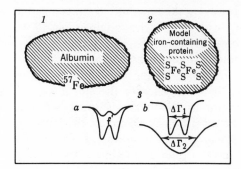

Figure 58. Monitoring of the local mobility of a protein matrix by nuclear γ resonance [15,16]: (*1*) Mössbauer labels (^{57}Fe) on the protein surface; (*2*) Mössbauer labels within the protein globule; (*3*) expected changes in the parameters of the γ-resonance spectra under conditions that allow local mobility: (*a*) harmonic vibrations, f' decreases; (*b*) diffusion broadening ($\Delta\Gamma$), f' remains constant.

Two types of ^{57}Fe-labeled proteins were used in the study: (a) albumin, in which the iron ions are bound by chelating groups of amino acid residues on the protein surface and (b) a model iron-containing protein, in which the iron ions are in a nonheme complex located within a protein fold [94]. The mobility of the iron atoms was determined from temperature dependences of the γ-resonance spectra of dry preparations, as well as frozen and liquid solutions in water and glycerol, and an aqueous suspension of labeled albumin. The state of the solutions and the surface layers of protein was monitored with the EPR spectra of spin probe R randomly distributed in the solutions and of spin label VII bound to the surface of the albumin macromolecule.

The measurements were carried out on an electrodynamic Mössbauer spectrometer with a radioactive ^{57}Co (Pd) source [318]. The temperature of the samples was varied continuously from 77° to 300°K. The resonance absorption spectra for all of the samples are well-defined doubtlets with quadrupole splitting and a chemical shift characteristic of the high-spin Fe^{3+}. The values of the chemical shift, splitting, and γ-resonance linewidth (in mm sec^{-1}) ($\delta = 0.73$, $\Delta = 0.8$, and $\Delta G = 1.6$) for all of the albumin and model iron-containing protein preparations remained constant in the range of temperatures investigated, indicating the constancy of the chemical bonds between the iron atoms and the proteins.

The analysis of the experimental spectra reduces to a determination of the probability of nonradiative resonance absorption (f') (from the area under the absorption curve) and the width of the doublet line at half the height. Figure 59 presents temperature dependences of the relative values of $f'_T/f'_{T \sim 190°}$ for the samples studied.

In the case of the dry protein preparations the f' value characteristic of solids is slightly dependent on the temperature, and the half-width of the γ-resonance line is constant in the 77–300°K temperature range. A different pattern is observed for the temperature dependence of f' in the case of the protein solutions. After varying slightly in the low-temperature region,

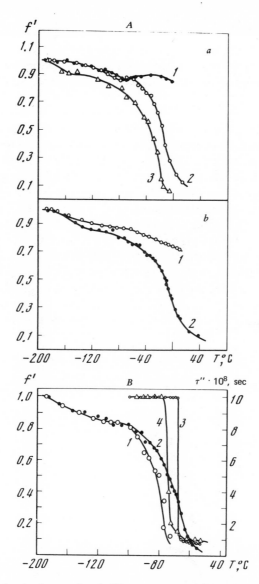

Figure 59. Variation in the rotational diffusion parameter τ_c'' of spin labels and probes and in the relative probability of the effect (f') without emission of Mössbauer labels (^{57}Fe) fixed to the protein matrices of HSA and the model iron-containing protein as functions of the temperature [15,16]: (A) f' for HSA (a) and the model iron-containing protein (b); (I) dry preparations; (2) glycerol medium; (3) HSA suspension; (b) f' in an aqueous medium for HSA (I) and the model iron-containing protein (2); (3) τ_c'' for probe R in the aqueous phase; (4) spin label I on the model iron-containing protein; the protein concentration is 100 mg/ml.

the value of f' decreases sharply at temperatures exceeding -60 to $-20°$ (Fig. 60). The broadening of the resonance absorption line in the protein solutions becomes significant only at low temperatures, where the probability of nonradiative absorption is small. As an example, Fig. 60 presents the γ-resonance spectra for solutions of albumin in glycerol. The observed broadening of the spectral lines does not exceed 50% of the linewidth at low temperatures. The decrease in the value of f' and the broadening of the γ-resonance line at temperatures ranging from 213 to 243°K, that is, under the conditions with which the spin probe in the solution remains maximally immobilized according to its EPR spectrum, are important factors. This implies that the ice matrix remains rigid even for a tiny probe; needless to say, this applies to the albumin macromolecule. Spin label VII is also allowed to move before the radicals in the bulk of the ice begin to move. The motion of the Mössbauer nuclei rigidly bound to the protein matrix, which is manifested as a decrease in the value of f' and broadening of the γ-resonance lines while the protein macromolecule is immobile, is naturally treated as a result of rapid local displacements of individual sections of the protein matrix relatively to the macromolecule that occur when small

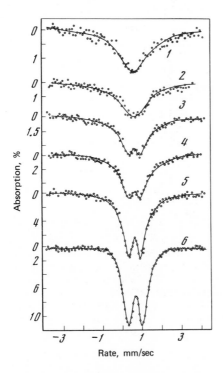

Figure 60. Diffusion broadening of the Mössbauer spectra of a glycerol solution of HSA (100 mg/ml) modified by ^{57}Fe [15,16]: (1) 13.5°; (2) $-4°$; (3) $-11°$; (4) $-22°$; (5) $-190°$; (6) $-196°$.

portions of the protein surface melt at temperatures lower than the melting point of the ice mass.

Analogous phenomena were observed in the study of the suspensions and viscous glycerol solutions. Thus in the samples of albumin modified with Mössbauer labels (the ^{57}Fe isotope), the γ-resonance method detects the following three types of motion:

1. High-frequency ($\omega \gg 10^7$ sec^{-1}) vibrations with small amplitudes ($A_2 < 0.1$ Å), showing a weak dependence of f' on the temperature (at $\leq -50°$ for solutions and $\leq +30°$ for dry preparations) and being similar to ordinary vibrational motions of nuclei in solids.

2. High-frequency vibrations ($\omega \gg 10^7$ sec^{-1}) with amplitudes A_2 of about 0.2–0.5 Å and a sharp dependence of f' on the temperature, which appear in the -40–$0°$ temperature range and intensify along the following sequence the model iron-containing protein, glycerol solution > the model iron-containing protein, aqueous solution > albumin, glycerol solution > the albumin suspension > albumin, aqueous solution.

3. Low-frequency displacements of the nuclei ($\omega \sim 10^7$ sec^{-1}), which are manifested by broadening of the γ-resonance line at temperatures of -10–$40°$ and intensify along the same sequence as in the preceding case.

The last two types of motion occur under the conditions for which the diffusion of the protein macromolecules is practically excluded (frozen solutions, albumin suspension, etc.) or its rate is several orders of magnitude lower than 10^7 sec^{-1}. Presumably, motions of nuclei with vibrational frequencies of about 10^7 sec^{-1} are due to spontaneous local conformational changes in the protein matrix that affect the polypeptide chain, since the iron ions, especially in the multinuclear complexes of the model iron-containing protein, are rigidly fixed on the polypeptide chains at many points. These changes are similar in nature to the local thermal variations in the structure of proteins that cause the mobility of snugly held spin probes (Chapter 7).

The results of the experiments with solutions of the model iron-containing protein are of special interest, since the iron–sulfur complex of the model protein is an analog of the active centers of natural iron–sulfur proteins [93,94].

The distinct dependence of f' and of the broadening of the γ-resonance line on the temperature for these proteins opens up new prospects for the direct study of the dynamic properties of the active centers of enzymes, if they contain Mössbauer nuclei. Gamma-resonance labels can be introduced into the active centers of enzymes, and the small dimensions of the γ-resonance labels in many cases makes them more convenient than spin labels.

ELECTRON-SCATTERING LABELS

Modern electron microscopy is a powerful tool for the study of biological structures [324–328]. The high resolving power of the best electron microscopes (1.5–3 Å) and the development of the methods of positive and negative contrasting in conjunction with biochemical techniques, makes it possible to solve numerous problems, including those associated with features of the tertiary and quaternary structures of proteins and enzymes. However, the electron microscope has still not been used to study the active centers of enzymes. This is due to the fact that carbon, nitrogen, oxygen, and hydrogen atoms contain a relatively small number of electrons (one to eight), and, since they are weak centers for the scattering of an electron beam, they do not produce clear electron-microscopic images. Biological structures with thicknesses less than 150–200 Å are usually made visible by shading them with layers of heavy metal atoms, such as tungsten, molybdenum, uranium, and lead, which contain a considerably higher number of scattering electrons (42–82).

However, the electron microscope can also be applied to the study of the active centers of enzymes and other biological systems by introducing electron-scattering labels, that is, compounds containing clusters of heavy atoms, into the structures under investigation. This technique has recently been applied in the Laboratory for the Kinetics of Enzyme Action at the Institute of Chemical Physics of the USSR Academy of Sciences to the study of the active centers of nitrogenase with the aid of labels containing heavy mercury atoms.

Research into the structure of the active center of nitrogenase with electron-scattering labels. The object studied was the nitrogen-fixing enzyme complex (nitrogenase) isolated from *Azotobacter vinelandii* according to a particular procedure [325] with several modifications. The molecular weight of the complex was 340,000 ± 60,000. According to certain data [191], the active center of nitrogenase contains nonheme iron and molybdenum. The application of inhibitors and spin labels [14,46] has shown that the iron atoms are bound to the protein globule by means of cysteine residues and are in close proximity to one another ($\leq$ 7–8 Å), the iron atoms being arranged in clusters rather than individually. However, the location of these clusters on the nitrogenase macromolecule and their relative arrangement were unknown.

The fact that the nonheme iron in nitrogenase is bound to the protein globule by means of cysteine residues makes it possible to use the electron microscope to determine the location of the iron, since the specific reagent for cyeteine groups, PCMB, contains the heavy mercury atom. After determining the location of the cysteine groups on the nitrogenase macromolecule from the position of the mercury replacing the iron, the original location of the nonheme iron can be revealed [17].

The method of negative contrasting [323] was used to study the macromolecular structure of the complex. The work was carried out on an Hitachi HU-125 electron microscope with an accelerating voltage of 100 kV and a working magnification of 100,000.

Figure 61 presents the original preparations of the native nitrogenase contrasted with phosphotungstic acid (PTA) in a 1:1 ratio. The photomicrographs show approximately round and rectangular molecules. The round molecules have a thin border of points, and their diameter is 100 ± 5 Å. The dimensions of the rectangular molecules are $100 \times 95 \pm 5$ Å. The unit cell of the molecules is a $40 \times 10 \pm 2$ Å tetrad consisting of spherical subunits, each of which is 20 ± 2 Å in diameter. The number of tetrads in the molecules is even, and in the rectangular forms it always equals eight. Among the round forms there are some incomplete complexes consisting of six tetrads.

When nitrogenase interacts with PCMB (see Fig. 61), partial disordering of the complex occurs. The characteristic packing of the subunits is disrupted, and the molecules lose some of their bulkiness and are spread over the background. Their diameter becomes equal to $\sim$ 120–130 Å. However, the dimensions of the subunits are not altered significantly, and the increase in the diameter of the molecules occurs as a result of the disruption of contacts between the subunits and the separation of the subunits by a certain distance.

This result implies that the iron in the nitrogenase molecules is somewhere close to the surface of the subunits and is most likely on the boundary between them. During the reaction with PCMB, in which the iron is displaced from the protein globule along with the sulfide ions, the contacts between the subunits are broken, and the molecule falls apart.

Since the mercury atoms have a high electron density, the nitrogenase preparations examined under the electron microscope were treated with PCMB, but were not contrasted with PTA. The resultant distinct electron photomicrographs (see Fig. 61) display the regular grouping of the mercury granules into clusters having the shape of tetrahedrons. The size of the granules is 6 ± 1 Å, and the distance between them is 8 ± 2 Å. According to the dimensions of the granules, we may assume that each corner of the clusters is formed by a minimum of four mercury atoms. This is supported by the fine structure of the granules. The data obtained suggest that the 32 iron atoms in the nitrogenase macromolecule are distributed in the following manner.

The iron is restricted to two unit cells of a complex consisting of eight subunits. In each subunit there are four iron atoms located in one portion

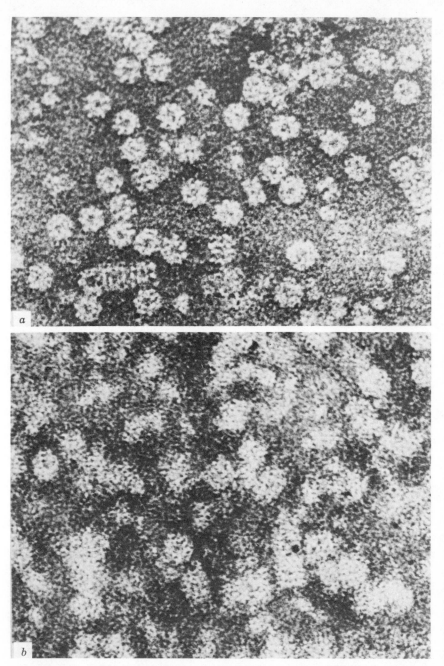

Figure 61. Electron photomicrograph of nitrogenase. Negative contrasting of phosphotungstic acid (PTA) [17]: (*a*) native nitrogenase; (*b*) nitrogenase treated with 100 equivalents of PCMB; (*c*) nitrogenase treated with 100 equivalents of PCMB without negative contrasting; (*d*) nitrogenase treated by 100 equivalents of PCMB with negative staining.

164

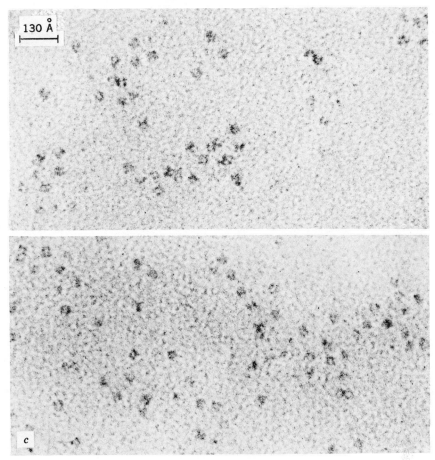

Figure 61. *Continued*

of the polypeptide chain, specifically at points of contact between the subunits. Since there are four subunits in one unit cell, the mercury cluster formed during the replacement of the iron by the mercurial has the form of a tetrahedron.

These results demonstrate the possibility of studying the structure of the active centers of enzymes with the aid of electron-scattering labels in those cases in which heavy atoms are concentrated in groups of four to six atoms. In the case of nitrogenase the presence of such groups is due to the multinuclear nature of the iron-containing center of the enzyme.

Further investigations [18] have shown that this approach can be extended to single functional groups of enzymes, if the label contains a group

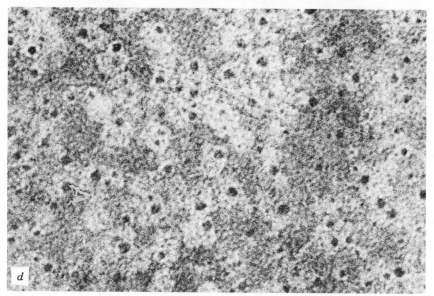

Figure 61.   *Continued*

of closely arranged heavy atoms. This requirement is fulfilled by the compound [329]

$$
\begin{array}{ccc}
\text{HO} & & \text{OH} \\
\diagdown & & \diagup \\
\text{Hg} & \text{Hg} & \\
| & | & \\
\text{Hg—C—C—Hg} & & \\
| & | & \\
\text{O—Hg} & \text{Hg—O} &
\end{array}
$$

The free SH groups of the model compounds of cysteine, glutathione, dithiothroetole, and bovine hemoglobin bind compound A in equivalent amounts. One molecule of compound A contains about 500 electrons and appears on the electron photomicrograph in the form of an individual point. In the case of nitrogenase labeled with A, pairs of closely arranged points appear in the field of view (Fig. 62). This implies that the free SH groups in the ATPase portion of the active center of the enzyme are near one another, since the binding of the labels is accompanied by cessation of the activity of the enzyme. Treatment of nitrogenase with compound A and then with PCMB results in the formation of closely arranged ensembles of mercury atoms (see Fig. 63). This result is naturally treated as a consequence of the close arrangement of the free SH groups in the iron-containing clusters of the nitrogenase active center. This conclusion is in good

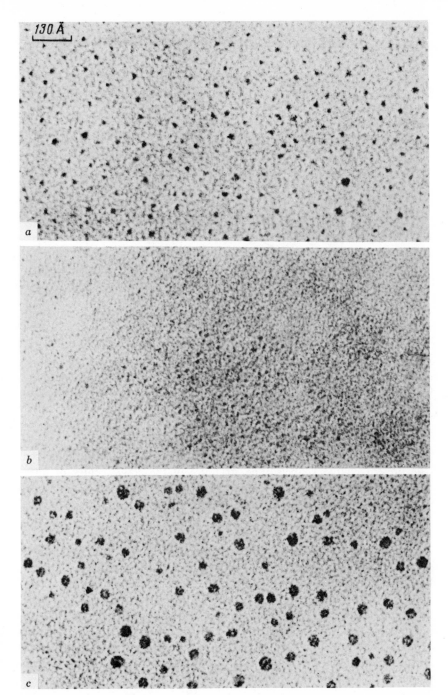

Figure 62. Electron photomicrographs of labels in the active center of nitrogenase [18]: (*1*) nitrogenase treated with label A (two equivalents); (*b*) label on a background layer; (*c*) label A on the SH groups of nitrogenase treated with 100 equivalents of PCMB.

167

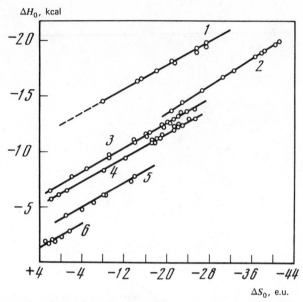

ΔH_0, kcal

ΔS_0, e.u.

Figure 63. Compensation dependences in the formation of myoglobin- and methemo-globin-ligand complexes at various pH [346]: (*1*) human hemoglobin, cyanide complex; (*2*) human hemoglobin, azide complex; (*3*) dog, pigeon, and guinea-pig hemoglobins, azide complex; (*4*) α chains of human hemoglobin and whale myoglobin, azide complex; (*5*) human hemoglobin, isocyanide complex; human hemoglobin, fluoride complex.

agreement with the data obtained with spin (Chapter 3) and luminescent (the present chapter) labels. The structure of the active center of nitrogenase is treated in detail in Chapter 5. The study of this enzyme is a good example of an investigation of a complex active center by combined methods.

A particular method proposed and developed [17,18] for introducing electron-scattering labels into the active centers of enzymes seems to be applicable to other classes of enzymes, including enzymes in complex biological structures. The application of mercurial derivatives capable of modifying various functional groups in proteins, nucleic acids, and other structures is also possible. The use of electron-scattering labels, like the luminescent and Mössbauer considered earlier in this chapter, considerably expands the arsenal of tools for the study of complex biological structures and is undoubtedly an addition to the traditional physicochemical techniques usually employed by biochemists.

Chapter Nine
Several Topics in Enzymology as Related to Data Obtained with Spin Labels

Introduction

Biological catalysts, that is, enzymes, effectively accelerate oxidation–reduction, acid–base, mechanochemical, and coupled reactions [331–365].

Chemical conversions are effected on enzymes by a large number of mechanisms. A detailed physicochemical investigation of specific enzymatic reactions is the main way of revealing the secrets of enzyme action. Spin labeling is a very effective and universal tool for studying the structure of active centers, the mechanisms for the action of enzymes, the nature and relative positions of catalytic binding groups, the topography of protein surfaces, and the local mobility of proteins (see Chapters 3–7). As shown in the study of many enzymes and especially in the case of α-chymotrypsin, different variants of the method make it possible to reveal fine-structural changes in active centers as they interact with substrates and their analogs. The main virtue of the method is thus the possibility of using it to investigate any specific enzyme.

No less significant is the prospect of using spin labels to study the general properties of enzymes. Presumably, the tremendous catalytic force and the highly specific nature of enzymes, as well as their superiority over most known chemical catalysts, regardless of the type of reaction being accelerated, reflect some general principles governing the structure and mechanisms of enzymes. If such principles do exist, they should certainly manifest themselves in the general nature of the physicochemical properties of enzymes.

The modern theories of enzyme action attribute the unique catalytic properties of enzymes to their multifunctional nature and dynamic structure. As shown in Chapters 3–7, in the case of the multinuclear iron-containing components of nitrogenase, spin labeling makes it possible to expand and define more precisely the concept of the multifunctional nature of enzymes, especially in the case of systems that cannot be studied by X-ray diffraction analysis.

The advantages of the method are vividly seen in the study of the dynamic structure of enzymes. The contemporary view of the dynamic structure of proteins is based on two hypotheses advanced about 15 years ago. According to the Linderström–Lang hypothesis [340], the structure of protein matrices is completely rigid, but has a dynamic character. The latter is due to local spontaneous changes in individual portions of the proteins. According to the Koshland hypothesis [334], the catalytic active structure of enzymes is not fixed beforehand, but is formed only on interaction with a substrate (induced fit).

It may be expected that the dynamic nature of the structure of enzymes should manifest itself in the following specific properties of protein macromolecules:

1. In changes in the conformation of the active center as it interacts with substrates, as well as in changes in the state of the protein matrix outside of the active center. These changes can be monitored with the aid of specific miniature seismographs, that is, spin labels added to the protein macromolecule.
2. In the optimum local mobility of individual portions of the water–protein matrix of enzymes due to the relative rigidity of the core and the softness of the outer shell of protein globules. The local mobility may be studied by introducing labels and probes into different layers of the matrix.
3. In the additional energy (ΔH_{conf}) and entropy (ΔS_{conf}) contributions of the conformational steps to the experimentally measured kinetic and thermodynamic parameters of enzymatic chemical processes.

In many cases spin labeling makes it possible to obtain information on the activation parameters for conformational changes and thereby yields material for comparison with the kinetic and thermodynamic laws governing enzymatic reactions.

The results of the investigation of allosteric and cooperative conformational changes in proteins and enzymes with spin labels are discussed in detail in Chapter 6, and questions regarding the rigidity and local mobility of protein macromolecules are treated in Chapter 7.

The present chapter deals mainly with energy and entropy aspects of conformational changes in proteins and enzymes. These changes manifest themselves in both the kinetics of enzymatic reactions and the rotational diffusion of spin labels and probes fixed to various portions of water–protein matrices (including the active centers of enzymes). In the concluding sections of the present chapter we present our view of the role of the multifunctional nature and dynamic structure of enzymes.

Energy and entropy features of enzymatic and denaturation processes. Several phenomenological laws. Kinetic methods make it possible to determine the experimental values of the parameters in the Arrhenius–Eyring kinetic equation

$$k = \frac{kT}{h} e^{(\Delta S^{\neq})/R} e^{(-\Delta H^{\neq})/RT} \tag{9.1}$$

and in the thermodynamic equation

$$\bar{k} = e^{(\Delta S)/R} e^{(-\Delta H)/RT}. \tag{9.2}$$

In applying Eqs. (9.1) and (9.2) to enzymatic reactions, it should be understood that because of the complexity and nonelementary nature of these processes, the parameters determined from temperature dependences of the rate constants for the steps in the conversion of the substrate (k_i), substrate-binding constants (k_s), inhibition constants (k_I), and so forth are effective parameters, which ultimately reflect the overall complexity of a given macrostep occurring on an enzyme (the presence of intermediates and externally undetected microsteps, the involvement of the protein matrix of the enzyme, and the effect of solvent particles). Therefore, an analysis of the kinetic and thermodynamic parameters of enzymatic reactions requires consideration of additional facts based on the theory of chemical kinetics, empirical data on simpler chemical and model systems, the physicochemical properties of enzymes, and so on.

The following three methods for analyzing kinetic and thermodynamic parameters of enzymatic reactions can be distinguished: (a) quantitative comparison of the energy and entropy parameters for enzymatic and model reactions, (b) theoretical evaluation of the parameters for different variants of the mechanism for the reactions under investigation, and (c) determination of correlations between the energy and entropy parameters of the enzymatic processes (the method based on the compensation effect).

The overwhelming majority of reactions catalyzed by enzymes have lower values for the effective activation energy than do the analogous uncatalyzed reactions [331,335]. The entropic parameters of enzymatic reactions take on extremely varied values, from -50 to $+50$ e.u. The entropy is often very sensitive to the structure of the substrate and enzyme, as well as the temperature and pH. Many cases of unusually high and unusually low $\Delta S^{\neq}$ and ΔS values, compared to the values of these quantities for simple model reactions, have been noted (Table 13).

Abrupt discontinuities are fairly often encountered on plots of log k_i and k as functions of $1/T$, indicating an instantaneous change in the effective energies and entropies of the reactions in a narrow temperature

Table 13.   Anomalous Entropic Parameters of the Steps of Substrate Conversion in Several Enzymatic Reactions

Enzyme	Substrate	$\Delta S^{\neq}$, e.u.	Reference
Chymotrypsin	Hydroxymethylhydrocyanate	−23.4	[376]
	of benzoyltyrosine	−21.4	
Acetylcholinesterase	Acetylcholine		
	aceylation	16—34	[377]
	deacylation	−42	
Trypsin	Hydrolysis of lactoglobulin		
	A	43	[378]
	B	55	
Pepsin	Carbobenzoyl-α-glutamyl-L-	−21.8	[379]
	tyrosine		
Fumarase	Fumarate, $t < 18°$	−23.4	[380]
Nitrogenase	ATP, N_2, $t < 20°$	45	[381]
Myosin	ITP	20.6	[382]
	ATP	22.7	
Lysozyme	Methyl-N-acetyl-β-D-glucosamide	−44	[383]
Horseradish ferroperoxidase	CO	23	[384]

range. In such cases the entropic parameters outside this temperature range generally have anomalous values (see Table 13).

The values of the enthalpy of formation of the enzyme-substrate and inhibitor complexes also vary within fairly broad limits, from −15 to +15 kcal.

The quantity $\Delta S_0^{\neq}$ for the elementary acts in simple gas-phase reactions equals 0–20 e.u. (in the case of monomolecular reactions) and −20 to −60 e.u. (in the case of bimolecular reactions not complicated by solvation between simple molecules).[1]

The values of the entropic parameters for gaseous chemical reactions usually fall well within the range of values calculated by statistical mechanics on the basis of the activated-complex theory and can be considered as a standard for "normal" elementary chemical processes.

According to Table 13, the values of $\Delta S^{\neq}$ for monomolecular enzymatic processes are outside of the range of "normal" values. The large negative values for such hydrolytic enzymes as chymotrypsin and pepsin are especially characteristic. Since the value of ΔS_M lies between −40 and −100 e.u., the positive experimental value of ΔS_M (Table 14) for many enzymes certainly indicates that there is a significant contribution from attendant processes in the binding of the substrate to the surface of the enzyme.

[1] The entropy loss on complexation (ΔS_H^0) equals −40 to −100 [373–375].

Table 14. Anomalous Entropic Parameters of the Steps in the Formation of Enzyme-Substrate and Enzyme-Inhibitor Complexes in Several Enzymatic Reactions

Enzyme	Substrate	ΔS_M, e.u.	Reference
Acetylcholinesterase	30 ligands of the type $R\text{--}CH_2\text{--}\overset{+}{N}\text{--}(CH_3)_3\cdot\overline{X}$	-4.3 to 20.2	[385]
α-Chymotrypsin	N-Benzoyl-L-alanine	10	[386]
Aspartate amino-transferase	α-Methylaspartate	-32	[387]
Ferricytochrome C	Imidazole	10.2	[388]
Myosin	ATP	49	[389]
Same	F Actin	234	[390]
Acetylcholinesterase	Acetylcholine	-14 to -29	[377]
	Aminoethyl acetate	7.5	
	Methylaminoethyl acetate	3.7	
Phosphate	m-Methoxyphenylphophoric acid	32.68	[391]
Fumarase	Fumarate, $t>18°$	26.8	[380]
Nitrogenase	ATP	17	[392]
	$Na_2S_2O_4$	19	
Liver microsomal hydroxylase	Octadecane	40	[393]

The following are tentative values for the enthalpy and entropy for the formation of bonds of various types corrected for the local dielectric constant in water; in the case of bonds between ions and dipoles, the value of ΔS was evaluated on the basis of desolvation of the ions and orientational effects [394,395]:

Type of bond	ΔH, kcal/mole	ΔS, e.u.
Covalent	-3 to -140	-30 to -50
Ion–ion	-3 to -15	$+6$ to $+20$
Ion–dipole	-0.5 to -4	-5 to $+5$
Ion-induced dipole	-0.5 to -2	-5 to $+5$
Dipole–dipole	-0.5 to -3	-5 to -25
Hydrogen	-1.5 to -12	-10 to -20
Hydrophobic	$+0.3$ to $+1.8$	$+1.5$ to $+12$

Correlation between energy and entropy parameters (compensation effect). In recent years another method has been developed for the quantitative analysis of the experimental data on the kinetics and thermodynamics of enzymatic processes [293,338,346]. As we have already noted, the present state of the theory does not permit calculation of the absolute values of

Table 15. Values of the Coefficients A_1 and β_1 in the Equation $\Delta H = A_1 + \beta_1 \Delta S$ (ΔH and ΔS refer to complexation processes)

Enzyme, Substrate	Variable Parameters	A_1, kcal	β_1, °K	β_1, T_{av}	Reference
α-Chymotrypsin N-Acetyl-3,5-di-bromo-L-tyrosine	pH	−2.8	270	0.92	[396]
α-Chymotrypsin	Substrate structure	−3.2	271	0.92	[346]
α-Chymotrypsin, indole	pH	−4.0	283	0.97	[379]
α-Chymotrypsin	Substrate and inhibitor structure	2.8	455	1.50	[399]
Acetylcholinesterase, 30 ligands of the type $R—CH_2—\overset{+}{N}—(CH_3)_3 \cdot \overline{X}$	Structure of R	−4.2	288	0.98	[385]
Ribonuclease, cytidine-3′-phosphate	pH	−5.5	276	0.96	[400]
Trypsin (17 substrates and inhibitors)	Substrate and inhibitor structure	−3.3	304	1.04	[401]
Phosphatase, derivatives of phenyl phosphoric acid	Substrate structure	−4.2	312	1.02	[391]
Methemoglobin C, thiocyanate	pH <7.5 >7.5	−4 −3.2	270 290	0.92 1.00	[346]
Methemoglobin A, thiocyanate	pH <7.0 >7.0	−3.8 −3.2	270 290	0.92 1.00	[346]
Methemoglobin C, cyanide	pH	−12.1	280	0.95	[346]
Methemoglobin A, cyanide	pH <7.5 >7.5	−11.3 −11.7	280 300	0.95 1.02	[346]
Methemoglobin A, fluoride	pH	−2.5	275	0.95	[346]
Methemoglobin C, fluoride	pH	−2.5	275	0.95	[346]
Chymotrypsin	Substrate structure	−1.4	300	1.00	[338]
Acetylcholinesterase	Same	−3.0	310	1.03	[338]
Carboxypeptidase	Same	4.0	300	1.00	[338]
α-Amylase	Activator anion structure	−5.4	290	0.96	[338]
Catalase	Inhibitor structure	−6.0	320	1.06	[338]
Agglutinin	Different agglutinins	−10.0	315	1.05	[338]
Fumarase	Substrate structure	4.0	290	0.96	[338]
Same	Inhibitor structure	2.0	290	0.97	[338]
Horse serum cholinesterase	Substrate structure	−5.0	270	0.93	[338]

Note. For equilibrium steps the values of ΔH and ΔS correspond to the standard enthalpies and entropies of complexation. In the other cases they are the effective values calculated from the temperature dependence of k_M.

Table 15 (end)

Enzyme	Variable Parameters	A, kcal	$\beta°$, K	β/T_{av}
α-Chymotrypsin	Substrate structure	22.0	420	1.4
Same	Same	17.0	420	1.4
Acetylcholinesterase (ACE)	Same	−21.2	520	1.6
Carboxypeptidase	Same	−24.0	720	2.4
Amylase	Activator ion structure	−17.5	230	0.8
Erythrocyte ACE	Reacted inhibitor structure	−13.5	260	0.9
Same	Amount of NaCl	—	270	0.9
Serum ACE	Enzyme structure	−11.5	460	1.4
h-Galactosidase	Substrate structure	—	280	0.9
D-amino-acid oxidase	Same	—	295	1.0

energies and entropies of activation for reactions in the condensed phase. There is, therefore, special interest in data not on individual enzymes and substrates, but on series of reactions in which specific parameters, such as the structure of the substrates and inhibitors and the reaction conditions (pH, solvent composition, and temperature) are varied on a regular basis. The main advantage in the new method is that it makes it possible to reveal the effect of individual factors while others remain approximately constant. For example, by varying the structure of the substituents in molecules with reaction centers of similar structure, we can isolate the components in the overall values of the thermodynamic parameters that are responsible for specific electronic or volume effects.[1]

The first paper in which a correlation was noted between the energy and entropy parameters of enzymatic processes was apparently that of Doherty and Vaslow [396], in which the values of ΔH and ΔS for the binding of the substrate N-acetyl-3,5-dibromo-L-tyrosine to α-chymotrypsin were determined at various pH values. The changes in pH were accompanied by parallel compensating changes in the ΔH and ΔS of binding.

In 1966 we analyzed the material available at that time on the kinetics and thermodynamics of enzymatic processes [338]. The analysis revealed some interesting laws governing the entropic and energy properties of many enzymatic reactions. Generally, in the case of one specific enzyme, an increase or decrease in the activation energy or enthalpy (ΔH) of a

---

[1] This refers to volume changes (ΔV) during the formation of the complex or the conversions. The value of ΔV can be determined from the dependence of the rate on the pressure [254,293] or by precision measurements [397,398].

process due to changes in the chemical structure of the substrate, inhibitor, or activator is accompanied by an increase or decrease in the entropy of activation or the entropy (ΔS), and the following relationships are fulfilled to an approximation:

$$E = A + \beta \, \Delta S^{\neq}, \tag{9.3}$$

$$\Delta H = A_1 + \beta_1 \, \Delta S. \tag{9.4}$$

The difference between the energy levels reaches 20 kcal, which, according to Eq. (9.1), should have caused changes in the rate constants or equilibrium constants of almost 15 orders of magnitude, but these changes are compensated by larger (≤ 60 e.u.) changes in the entropies, and the resultant differences in the actual constants are considerably smaller. Dependences such as Eqs. (9.3) and (9.4) have been termed compensation effects. The values of the coefficients β_1 in equations such as (9.4) are similar to the average temperatures of the experiments (T_{av}) (Tables 15 and 16). The same relationship is also fulfilled in experiments with a particular enzyme in different temperature ranges. In the case of kinetic parameters corresponding to different substrates, activators, or inhibitors β is generally not equal to T_{av}.

In a number of cases, for example, for carboxypeptidase and α-chymotrypsin, the observed correlation is only approximate. However, in other experiments a linear relationship has been established with a fair degree of reliability between the energy and entropy parameters.

The review by Lumry and Rajender [346] presents additional material on the compensation effect in biological processes. The most significant examples in which the thermodynamic parameters for substrate–enzyme interactions were recorded with a high level of accuracy were the processes involved in the formation of complexes between hemoglobin and cyanide, azide, thiocyanate, and fluoride at different pH values and in the binding of the hydrocyanamate of N-acetyltryptophan to chymotrypsin (Fig. 63). All these processes are characterized by values of the coefficient β equal to 270 to 290°K, which are close to the experimental temperatures. The well established correlation between the energy and entropy parameters for the reactions of α-chymotrypsin is shown in Fig. 64.

Processes involved in the denaturation of proteins. The rate of the destruction of the tertiary and secondary structures of proteins during denaturation under the effect of elevated temperature, acid and alkaline pH, and various reagents can be measured with a fair degree of accuracy from the variation in the degree of spiralization in the molecules, the optical density of the solutions, the increase in the concentration of the coagulating

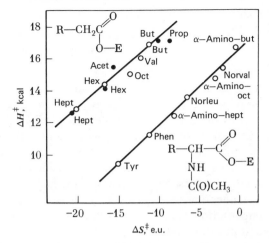

Figure 64. Compensation effect in the deacylation of acylated α-chymotrypsin derivatives [371].

fraction, and the loss of enzymatic activity. Although in each specific case the term denaturation requires a suitable explanation, it is usually assumed that the denaturation of proteins involves conformational changes of the globule–random coil type in large portions of the macromolecules. The main kinetic features of denaturation processes are as follows:

1. There are abrupt changes in the denaturation rates in a small temperature range and in a narrow range of concentrations for hydrogen ions, organic solvents, and certain other compounds.

2. If the experimental data can be expressed in terms of the Arrhenius–Eyring equation, the denaturation process may be characterized by unusually large (in comparison to the parameters of ordinary chemical reactions) values of E_{eff} and $\Delta S_{eff}^{\neq}$, in some cases reaching 180 kcal and 600 e.u. per mole of protein.

3. The values of the parameters E_{eff} and $\Delta S_{eff}^{\neq}$ are extremely sensitive to changes in the ambient conditions. In particular, the denaturation kinetics are substantially dependent on the aqueous environment of the proteins. Moistening dry preparations causes a significant increase in the E_{eff} and $\Delta S_{eff}^{\neq}$ of denaturation [402,403]. Replacement of water by deuterated water also causes significant changes in the kinetic parameters of denaturation [403].

4. There is a linear relationship (compensation effect) between the experimentally determined values of the activation energies E_{eff} and entropies of activation ($\Delta S_{eff}^{\neq}$) corresponding to different proteins or to the same

protein under different conditions. This relationship is expressed by the equation [414,415]

$$E_{eff} = \alpha + \beta \, \Delta S_{eff}^{\neq}. \tag{9.5}$$

When we go from one protein to another or when the pH or the concentrations of denaturing agents (alcohol, salts, etc.) are varied, the energy and entropy of activation almost always vary in parallel. The slope of the straight in E versus $\Delta S^{\neq}$ coordinates corresponds to the slope of the straight line for the values of the enthalpy (ΔH) and entropy (ΔS) of reversible denaturation. Kinetic and thermodynamic compensation effects are thus characterized by similar values of the coefficient β in Eqs. (9.3) and (9.4) and identical scales for the changes in the energy and entropy parameters. These phenomena seem to be caused by the same physical factors. The compensation effect in denaturation is considerably greater than that observed in other systems.

According to the existing theories, the processes involved in the denaturation of biopolymers are cooperative conformational transitions. The thermodynamic theory of cooperative systems as applied to biopolymers has been developed in various studies [405–408].

All of the foregoing features of the denaturation of biopolymers are explained on the basis of models that take into account the features of the proteins as cooperative systems [404,407].

Denaturation is often considered as a "melting" process. However, in the case of proteins we can only speak of approximate analogies to melting and phase transitions. Proteins are "aperiodic" irregular crystals. The energy of cooperative bonding differs from one part of the molecule to another, and places where the cooperative bonding is disrupted, defects, and so forth are encountered. There are parts of a system with properties approximating amorphous viscous liquids. All of these properties, which are present to different degrees in different proteins, cause the changes in proteins to be somewhat blurred or diffuse. Actually the cooperative processes in proteins are intermediate between ordinary chemical reactions, in which a small number of molecules is involved, and large-scale cooperative phase transitions involving a tremendous number of identical particles.

Kinetic models of conformational changes in proteins.   Even the first investigations of inactivation and denaturation of proteins revealed unusual kinetic effects. For example, the dependence of the rate constants on the temperature is only formally described by an Arrhenius equation. The enormous values of the activation energies in the equation cannot be explained by the theories of active collisions.

Eyring [409] developed the theory of an activated complex and theorized that the denaturation of proteins is accompanied by the simultaneous cleavage of a large number of bonds, which in the transition state make a contribution to both the enthalpy and entropy of activation. The positive value of the entropy of decomposition formally explains the fact that the process proceeds with a measurable rate despite the large activation energy.

In 1961 Vol'kenshtein and his associates [410] and subsequently Gotlib [411] carried out theoretical investigations of the kinetics of cooperative processes involved in structural changes in a system with two or more discrete values for the parameters of state. The probability of a change in state in each subsystem was assumed to be dependent on the state of the neighboring subsystems. The important result of these investigations is the conclusion that there is a sharp change in the rate of the process when a critical temperature (T_c) is passed.

In 1964 we [415] advanced the hypothesis that the experimentally observed kinetic laws governing the denaturation of biopolymers are due to the cooperative nature of the processes. A model based on the mathematical apparatus of Markovian chains was proposed for a phenomenological description of these processes.

In 1965 the same workers [414] and somewhat later Koshland and his co-workers [412] analyzed similar models of induced conformational changes in macromolecules involving a number of intermediate states. It was assumed [414] that during a conformational change in a macromolecule a certain monomer passes irreversibly or reversibly from the initial state to the final state (transition induction), and this change causes a chain of transformations in the neighboring segments. Analysis of the models makes it possible to account qualitatively for several features of the kinetics of biopolymers, including the compensation effect.

The Koshland model [412], which is based on the idea of an induced correspondence between the conformation of an enzyme and the structure of the substrate, assumes that changes in the conformation of enzymes under the effect of substrates occur by means of a sequence of several intermediate states from the initial state to the final state. As in the preceding model, each new state arises on the boundary of a segment where a transition has already occurred.

In that same year Vol'kenshtein proposed a method for analyzing cooperative kinetic effects in multicentric enzymes, which suggested that the kinetic parameters governing the functioning of an active center depend on the state of the neighboring centers [413].

Thus the authors of the papers just cited all start out from similar assumptions, the principal one being the assumption that the thermodynamic or kinetic parameters of centers depend on the state of the neighboring centers. A model based on the following assumptions was considered in various studies [288–293], namely, that there is a system consisting of a large number of identical ensembles each of which consists of n parts (links) joined to one another. When n is large, the transition of the system into a certain final state cannot occur within the time of a single elementary act (10^{-13} sec). Therefore, according to the most likely mechanism for the process, a transition occurs in the system in a small number of links (induction) with the energies and entropies of activation E_i and $\Delta S_i^{\neq}$. The weakening of the cooperative interaction in this segment causes the subsequent elementary transitions to become faster. The calculations showed that all of the basic laws governing conformational changes, including the compensation effect, are readily explained.

In particular, the experimental values of E_{eff} and $\Delta S_{eff}^{\neq}$ must be linearly dependent on the number of links in a cooperative segment if these links undergo changes at a temperature below the "melting point" of the cooperative ensemble

$$T_c = \frac{\Delta H_0}{\Delta S_0},$$

where ΔH_0 and ΔS_0 are the enthalpy and entropy of one elementary change, respectively. Here we have

$$E_{eff} = E_i + n \cdot \Delta H_0, \tag{9.6}$$

$$\Delta S_{eff}^{\neq} = \Delta S_i^{\neq} + n \cdot \Delta S_0. \tag{9.7}$$

Above the "melting point" we have $E_{eff} \sim E_i$ and $\Delta S_{eff}^{\neq} \sim \Delta S_i^{\neq}$. Thus the value of E_{eff} must change in the vicinity of the melting point (T_c). In the case of systems differing with respect to the size of the cooperative segment, compensation equation (9.3) must be obeyed with

$$A_1 = E_i - T \Delta S_i^{\neq} \quad \text{and} \quad \beta = \frac{\Delta H_0}{\Delta S_0} = T_c.$$

It is interesting that, according to the theory, curves in Arrhenius coordinates with different n need not intersect at one point ($T = \beta = T_c$), as predicted formally by Eq. (9.5) [416,417], but must have a discontinuity, whose abruptness increases with n. Moreover, there must be no conversion in the reaction series [417].

These conclusions were tested experimentally in the case of the denaturation of the enzyme catalase [290]. Measurement of the rates of the inactivation of catalase (which vary over almost seven orders of magnitude), with the aid of specially developed methods (thermoadiabatic measurements and determinations of residual concentrations) at temperatures ranging from 40 to 90°, showed that there is a compensation effect in catalase samples with different degrees of structural disruption, that is, with different n, in agreement with the theoretical assumptions. In the region of the isokinetic temperature there is a change in E_{eff} from 96 kcal in the low-temperature region to 45 kcal in the high-temperature region.

The data on the compensation effect in enzymatic reactions can be explained in an analogous manner, if we assume that the experimental energy and entropy parameters include added on conformational terms whose values are dependent on the size of the cooperative segment (n). Similar theories explain the data on the kinetics of enzyme action by the high sensitivity of n to various effects.

DYNAMIC STRUCTURE
AND CATALYTIC PROPERTIES OF ENZYMES

The main arguments in favor of the dynamic conception of proteins can be summarized as follows:

1. The energy and entropy laws governing enzymatic reactions, including compensation effects and entropic anomalies, are readily explained in the framework of models that assume reversible changes in the conformation of the protein or the structure of the environment during enzymatic processes.
2. There is a similarity between the enthalpy–entropy correlations (compensation effects) in the chemical steps of enzymatic reactions and "conformational" processes such as the local conformational changes monitored with the aid of spin labels and denaturation processes (see Fig. 53). However, the order of magnitude of the compensation effect in chemical processes is much lower than in denaturation, apparently because of the limited nature of the conformational changes that can occur in enzymes during normal functioning.
3. The results of the study of the state of various portions of protein matrices with the aid of spin labels together with the data on the hydrodynamic properties [408] and hydrogen exchange [418,419] in proteins imply that in the native state proteins have a compact and rigid core and a softer and conformationally mobile outer shell. The local viscosity of water at a distance of three or four layers from protein surfaces is close to the viscosity of pure water. Such a combination of properties seems optimal for the limited mobility of the individual segments of enzymes when the basic configuration is maintained.
4. In the case of many enzymes and electron carriers (lysozyme, myoglobin, hemoglobin, aspartate aminotransferase, and myosin) it is possible to detect conformational changes in the vicinity of spin labels at distances of ≥ 15 Å from the active centers during the specific complexation of enzymes with substrates and inhibitors.

The present theoretical ideas concerning the role of the dynamic structure in bringing about the high catalytic activity of enzymes can be divided into two groups. According to the theories in the first group, the fundamental laws governing enzymatic activity can be reduced to the conventional physicochemical laws governing nonbiological chemical reactions [193,334,335,346,347,352,365–372]. According to the other group of theories, enzymes possess a qualitatively new set of properties that make them more like mechanical devices [338,342,349,350,361,363,364]. For example, enzyme molecules are said to have the ability to utilize the energy of elementary chemical reactions by converting it either into conformational energy of the protein or into mechanical or acoustical vibrations of the macromolecules. In subsequent acts the stored energy is used to reduce energy barriers in the steps of chemical processes that are most difficult from the energy standpoint [338,342,349].

The principal difference between these two groups of theoretical ideas is that the first group treats the concept of an energy and entropy of activation in the context of chemical kinetics and permits the use of the theory of the activated complex, while, according to the ideas of the second group, the concept of activation parameters is devoid of physical meaning.

A more detailed treatment of the theories of enzymatic activity is beyond the scope of this book. We note only that, from our point of view, the physical mechanisms for the utilization of energy are less substantiated than are the chemical theories of enzymatic activity. The latter rest on a solid foundation of specific experimental data on enzymatic reactions and take into account the results of investigations in model chemical systems.

Enzymes, the optimal catalysts.   Before we discuss the role of conformational changes in enzymatic chemical reactions, let us consider what properties a certain ideal optimal catalyst [420] must possess according to the theories of modern chemical kinetics.

The rate of a catalyzed chemical reaction is largely determined by its free-energy profile. The presence of deep valleys or high plateaus on the path of the free energy changes along the reaction coordinate can slow the process considerably. Therefore the main function of catalysts is to evenly distribute the energy among the steps in such a way that intermediate compounds with specific structures are formed [50,81,370,420,422]. According to the law of the conservation of energy, the value of the free energy of the overall effect (ΔF_0) and the free energies of the individual steps must obey the unambiguous relation $\Delta F_0 = \sum_i \Delta F_i$. This means that, when the value of ΔF_0 is fixed, excessive favoring of some intermediate steps automatically makes others more difficult; therefore, the optimal catalyst is a catalyst with a maximally smooth energy relief in which $\Delta F_1 = \Delta F_2 = \ldots = \Delta F_n = (\Delta F_0/n)$.

It is known [373–375] that a very precise mutual orientation of the reacting molecules is necessary for an effective chemical conversion. Such an orientation usually requires a significant loss of translational and rotational degrees of freedom, which causes a decrease in the rate constants of the processes by a factor equal to 10^{-4}–10^{-8} (steric factors and negative entropies of activation). The rate constant can be increased by a factor approximately equal to its reciprocal, if the optimal catalyst guarantees the precise orientation of the reactant groups beforehand by decreasing the enthalpy (ΔH) of the processes on formation of a complex with contact binding groups of the catalyst.

It can be shown that the effective activation energy of the process is thereby decreased by ΔH. In other words, the catalyst uses the enthalpy of complexation to facilitate the difficult step in the conversion of the substrate by slowing the easy step of product desorption, which has no harmful effect on the overall process. However, it is important that the value of ΔH not be too large a negative quantity. Acceleration due to a mechanism of prior substrate orientation can take place only if the bond between the orienting and oriented groups is not broken in the activated state. This requirement is not usual or automatically fulfilled in all systems. Any given chemical conversion is accompanied by changes in the distance between the reacting nuclei by at least 1–2 Å (the difference between the van der Waals and chemical diameters of atoms). Separation of the orientating and oriented groups by such a distance must result in almost complete cleavage of the bonds between them.

Therefore a necessary condition for a gain in rate due to orientation is that some substrate nuclei are mobile while others are not. This may be achieved by altering the shape of the catalyst during the reaction without a significant loss of contacts between the substrate and the binding groups (dynamic adaptation).

All of our experience with chemical kinetics indicates that there is a significant correlation between the kinetic and thermodynamic parameters of different types of reaction according to a "the better the thermodynamics, the better the kinetics" principle. Examples of this correlation are: (a) the proportional relationship between the activation energy and the enthalpy of radical processes in the gaseous and liquid phases (the Polanyi–Semenov rule), (b) the similarity between the values of the activation energies and the bond energies in monomolecular processes, (c) the relationship between the rate constants of ion and ion–molecule reactions and the thermodynamic parameters of certain standard processes (the rule of Bronsted and Hammett), and (d) the linear relationship between the logarithms of the rate constants for electron transfer and the free energies of transfer. Correlations of this kind have also been observed in chemical catalysis [428–431]. It may, therefore, be expected that the optimal catalyst will show a higher rate for a thermodynamically more favorable elementary step. For this reason a good catalyst must carry out the process with the least number of intermediate steps. Each new step produces an additional residence time. According to the principle of "the better the thermodynamics, the better the kinetics," one- or two-step processes are the most favorable in the case of exothermal and thermoneutral reactions. Multinuclear and multifunctional catalysts, which are capable of simultaneously (synchronously) attacking different portions of the substrates, are a suitable apparatus for bringing about energetically favorable processes. Thus an optimal catalyst must make possible the synchronous action of a concerted mechanism to reduce the activation energy by taking advantage of its multifunctional nature. It is reasonable to suppose, however, that there are definite kinetic restrictions for effecting multinuclear synchronous mechanisms. The effective thermal motion of the nuclei along the reaction coordinate requires spatial and time synchronization. The statistical and thermal nature of chemical processes certainly limits the number of atomic nuclei participating in an elementary act to the lifetime of the transition state (10^{-13} sec).

Thus a minimum number of nuclei should be displaced in an elementary step (the principle of minimal motion) [432–435].

However, the action of restricted synchronous mechanisms that bring about a transition into an activated state without marked changes in the original structure can be visualized. Favorable kinetic and thermodynamic factors could be combined together in such processes.

According to the data in two studies [81,293], an optimal catalyst must provide for motion along the reaction coordinate of a minimum number of nuclei in each step, but this must occur without loss of the necessary contacts between the substrate and the catalytic groups in the transition state (the principle of optimal motion). The preservation of these contacts can be guaranteed by the optimal mobility of the matrix, which supports the catalytic and binding groups. A direct consequence of this principle is the multistep nature of catalytic reactions, as well as the dynamic correspondence between the shape of the catalyst to the successively changing structure of the reagents and catalytic groups.

Finally, processes occurring on optimal catalysts must not include quantum-mechanically forbidden steps that involve changes in the total electron spin and disruption of the symmetry and molecular orbitals [436–441].

Analysis of the data on the structure and mechanism of the action of enzymes implies that their overall properties are very similar to the properties of an optimal catalyst. First of all, they offer a wide assortment of physicochemical mechanisms for distributing energy among such steps as:

(a) the formation of intermediate compounds held together by a covalent bond, (b) acid–base catalysis, (c) electrostatic (ion–ion and ion–dipole) interactions, (d) the unique pattern of the local dielectric constant in the active centers of enzymes, and (e) interactions in the coordination sphere of metals, including the formation of multinuclear complexes in enzymes that contain several metal atoms in each macromolecule, which are considered to play a special role in the optimization of the energy profile of enzymatic reactions [428].

Enzymes, like globular proteins, are formed by very complicated packing of the polypeptide residues and have a large number of hydrophobic, electrostatic, and hydrogen contacts. According to the ideas developed by Lumry [360], changes in the conformation of proteins on formation and conversion of enzyme-substrate complexes result in the disruption of some contacts and the formation of others and thereby make a certain contribution to the thermodynamic parameters of the elementary steps. It may be stated that proteins, being cooperative systems, have large energy and entropy "capacities," which can be used to transfer the energy of the steps and to level off the energy profile.

Proteins with their unique possibilities of arranging the catalytic groups at different distances from one another are capable of forming stressed rings with substrates and of compressing or stretching substrate groups.

The distribution of the energy among the steps of an enzymatic process can be established by three groups of methods: (a) experimental study of the fast intermediate steps, (b) chemical simulation of the individual steps in a process, and (c) theoretical analysis of the energetics of hypothetical intermediate steps.

Despite the great difficulties in carrying out the experiments, the first two methods have been successfully applied to the study of many reactions, of which we note those involving α-chymotrypsin [443], aspartate aminotransferase [345], catalase [442], and D-amino-acid oxidase [444, 446]. In almost all cases studied, we can say that there is an even distribution of the free energy among the steps and that the energy barriers for the individual steps are not high.

As shown in Fig. 65, the existing data on enzymatic reactions does not contradict the principle of "the better the thermodynamic, the better the kinetics." There is an approximate linear correlation between ΔF and $\Delta F^{\neq}$ for the steps of enzymatic processes (the values of $\Delta F^{\neq}$ were calculated with Eyring's formula). Obviously the number of reliable examples known is small, and final conclusions cannot yet be drawn. Nevertheless, enzymatic reactions for which the kinetic and thermodynamic parameters have been measured can be divided into the following three groups: (a) reactions that evolve heat (the catalase reaction, $k = 5 \times 10^7 \ M^{-1} \cdot sec^{-1}$), (b) reac-

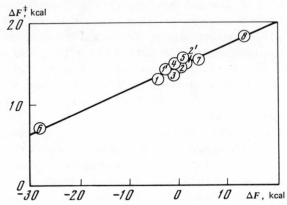

Figure 65. Correlation between the free energies (ΔF) and free energies of activation ($\Delta F^{\ddagger}$) of enzymatic reactions: (*1,1'*) fumarase, various steps [380]; (*2,2'*) AAT, various steps [330]; (*3*) acetylcholinesterase, overall reaction [255]; (*4,5*) α-chymotrypsin, various steps [346]; (*6*) catalase [442]; (*7*) alcohol dehydrogenase, pH 7; (*8*) nitrogenase, hypothetical step involving the reduction of molecular nitrogen to hydrazine accompanied by the hydrolysis of one molecule of ATP [50,81].

tions for which the free energy is close to zero and turnover rates are approximately 200–500 sec^{-1}, and (c) the slowest of all, the reduction of molecular nitrogen (about one turnover per second), whose rate-limiting step is, in all likelihood, the conversion of nitrogen into a hydrazine derivative [50], which is energetically unfavorable.

Although the existence of definite synchronous mechanisms has been assumed in the case of many enzymatic processes, at the present time there are very few concrete proofs that this is correct. However, we can cite at least one enzymatic reaction that proceeds via a synchronous mechanism. This process is the fixation of molecular nitrogen involving the enzyme nitrogenase under the effect of biological and model substrates whose reduction potentials are close to the reduction potential of molecular hydrogen. According to the analysis of the thermodynamics of various mechanisms for the reduction of N_2 in [50,81,293], one- and two-electron (atomic) mechanisms must be characterized by high activation-energy values, close to 90 and 40 kcal, respectively, and, therefore, cannot be activated under mild conditions without an additional outlay of energy. Multielectron reduction mechanisms are either exothermic or thermoneutral and are not assumed to involve the surmounting of high energy barriers. However, these mechanisms call on at least two atoms of the catalyst that are capable of transferring at least four electrons (along with the respective protons) in one act and are thus similar to synchronous push–pull mechanisms.

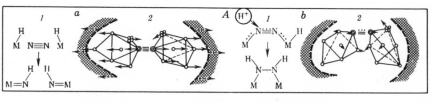

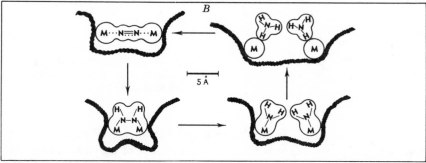

Figure 66. Mechanisms for the reduction of molecular nitrogen in the active centers of nitrogenase [50,293]: (*A*) six-electron (imide) (*a*) and four-electron (hydrazine) (*b*) mechanisms for the reduction of N_2; (*1*) chemical reactions; (*2*) schematic representation of the motion of the nitrogen nuclei, the central atoms, the ligands, and the protein groups based on the probable values of the covalent and van der Waals radii; (*B*) schematic representation of the changes in the configuration of the reagent, the catalytic groups, and the protein matrix during the reduction of molecular nitrogen to ammonia based on the covalent and van der Waals radii in the respective complexes (dynamic adaptation).

The nitrogenase reaction can also be used to illustrate the principle of optimal motion of nuclei in the electron act of enzymatic processes [293, 365]. As seen from Fig. 66*a*, the reduction of N_2 by a six-electron mechanism requires the simultaneous displacement of nitrogen, metal, ligand, and protein nuclei and is, therefore, unlikely. In a limited synchronous four-electron mechanism for the reduction of N_2 in hydrazine, only a minimal number of nuclei are displaced, and the energetics of the process are fairly favorable [293]. However, this forces us to assume that there are several more steps that require maintenance of the active contacts in all steps of the complex process by changing the conformation of the protein matrix (see Fig. 66).

Consecutive fast reversible conformational changes can provide for the effective restructuring of proteins without substantial expenditures of conformational free energy ($\Delta F_{conf} = \Delta H_{conf} - T \Delta S_{conf} \sim 0$). Such restructuring must be accompanied by simultaneous changes both in the enthalpy and the entropy of the system. It is, therefore, not surprising that compensation effects are fairly often encountered in biological systems.

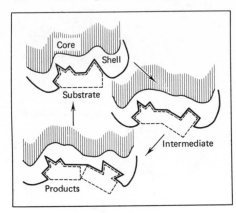

Figure 67. Schematic representation of the changes in the configuration of the substrate and the protein matrix in a consecutive reaction on an enzyme (dynamic adaptation).

If we assume that the motion of spin and Mössbauer labels and probes in protein macromolecules and in the active centers of enzymes reflects the corresponding spontaneous conformational changes in the protein matrix, the frequency of these changes can be estimated at 10^6–10^7 sec^{-1}. Since this value exceeds the turnover number of enzymatic processes, limited conformational changes can provide for the dynamic fit of the shape of an enzyme to the successively changing structure of the reaction groups of the substrate and enzyme in each step of the process. These considerations are illustrated in Fig. 67. According to this figure, thanks to the easy transition between enzymes conformations I, II, and III, which have practically equivalent free energies, the protein matrix undergoes dynamic adaptation during the entire process, thus promoting both the first step of precision orientation and the subsequent chemical steps without allowing cleavage of the contacts that are needed for the chemical mechanism.

As already noted, the ability of proteins to change their conformation is also very important for the creation of a favorable energy profile for enzymatic processes by optimally distributing the energy and entropy between the substrate and the enzyme [346,360].

Thus, in our opinion, the dynamic adaptation and even distribution of energy among the steps are the main functions of the conformational changes and the dynamic structure of enzymes. We would like to advance the following hierarchic relationship between the functional behavior of enzymes (their specific nature and high catalytic activity) and the structural and dynamic properties of proteins:

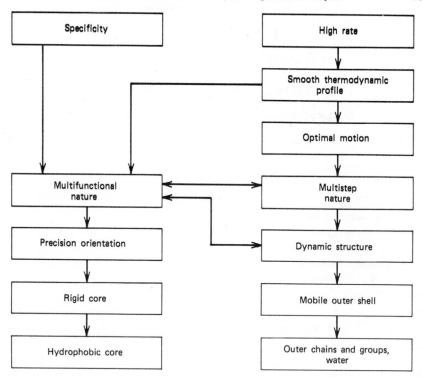

This interrelationship, which has been nearly perfected by biological evolution, accounts for the functioning of enzymes under conditions that approximate those for the functioning of the optimal catalyst.

Chapter Ten
Application of Spin Probes to the Study of Various Biological Systems

BIOLOGICAL MEMBRANES

Even the first attempts to employ nitroxide spin probes in the study of biological membranes [48,51,450,451] demonstrated the prospects of this new technique. The density, mobility, and polarity of the local environment of nitroxide radicals in a membrane can be evaluated from their EPR spectra. Information can be obtained on the packing, orientation, and rotational anisotropy of membrane lipids. Unique information on the lateral diffusion and flip-flop transitions of lipids in bilayer membranes has been obtained with the aid of spin probes. Membrane proteins and protein–lipid interactions in membranes have been successfully investigated. The study of structural changes in natural and model membranes induced by various factors has been the subject of a large number of papers. Attempts have recently been made to use nitroxide radicals to evaluate the fraction of membrane lipids that become part of the lipid bilayer structure and to determine the dimensions of these structures. Nitroxide radicals are effective electron acceptors and can, therefore, serve as redox probes to study electronic donor–acceptor interactions in biological systems.

This chapter presents several features of the spin-probe method that should be kept in mind in investigations of biological membranes with such probes. In addition, examples of the application of the method to the study of the structure and functions of biological membranes are given.

Features of the EPR spectra of spin probes in biological membranes. The most effective spin probes used in membrane investigations are spin-labeled derivatives of fatty acids (LIV–LVII), steroids (LX–LXI), and lipids (LXII–LXIII), which bind specifically to the hydrocarbon segments of membrane lipids (see below).

$C_7H_{15}COO$— N—$O^{\cdot}$ $C_{15}H_{31}COO(CH_2)_2$— N—$O^{\cdot}$

LII   LIII   LIV   LV

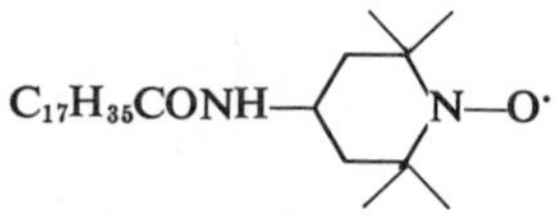

$C_{17}H_{35}CONH$—[ring]—N—$O^{\cdot}$

LVI

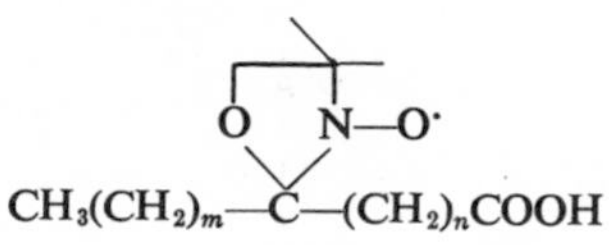

$CH_3(CH_2)_m$—$\overset{\overset{\displaystyle O\quad N-O^{\cdot}}{|}}{C}$—$(CH_2)_n COOH$

LVII

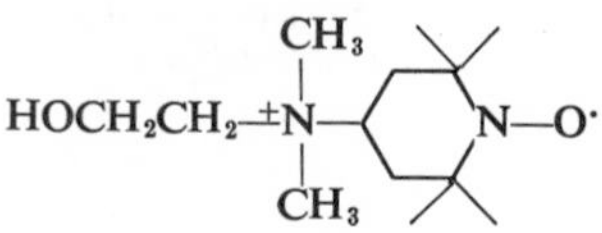

$HOCH_2CH_2$—$\overset{\overset{\displaystyle CH_3}{|}}{\underset{\underset{\displaystyle CH_3}{|}}{\pm N}}$—[ring]—$N$—$O^{\cdot}$

LVIII

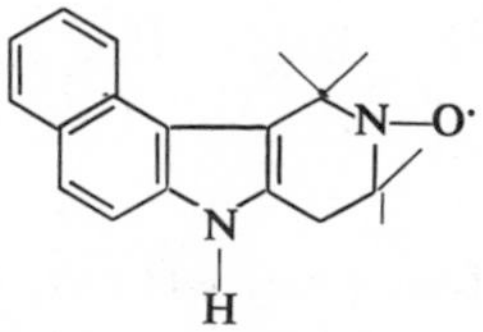

LIX

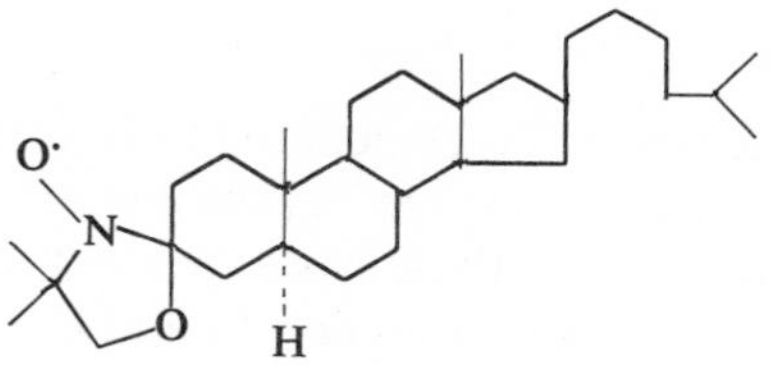

LX

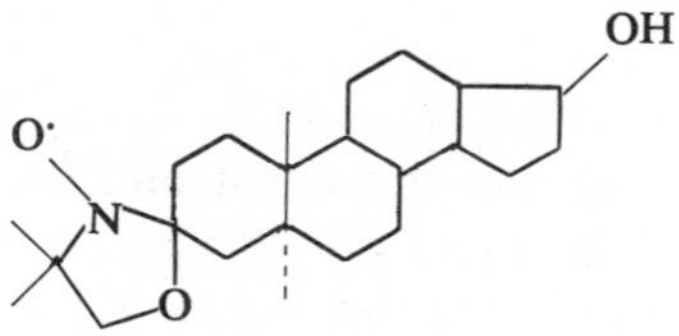

LXI

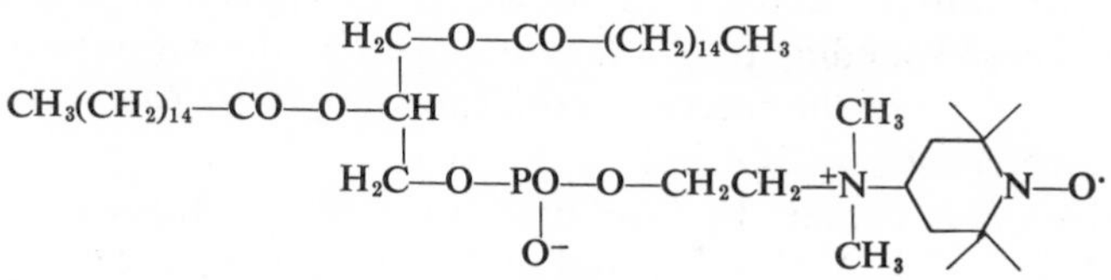

H_2C—O—CO—$(CH_2)_{14}CH_3$

$CH_3(CH_2)_{14}$—CO—O—$\overset{|}{C}H$

H_2C—O—$\overset{\overset{\displaystyle }{|}}{\underset{\underset{\displaystyle O^-}{|}}{P}O}$—$O$—$CH_2CH_2$—$\overset{\overset{\displaystyle CH_3}{|}}{\underset{\underset{\displaystyle CH_3}{|}}{\pm N}}$—[ring]—$N$—$O^{\cdot}$

LXII

$CH_3(CH_2)_m$—$\overset{\overset{\displaystyle O\quad N-O^{\cdot}}{|}}{C}$—$(CH_2)_n$—$\overset{\overset{\displaystyle O}{\|}}{C}$—$O$—$\overset{|}{C}$

H_2C—O—C—$(CH_2)_{\overline{m+n+1}}CH_3$

H_2C—O—$\overset{\overset{\displaystyle }{|}}{\underset{\underset{\displaystyle O^-}{|}}{P}}$—$O$—$(CH_2)_2$—$\pm N(CH_3)_3$

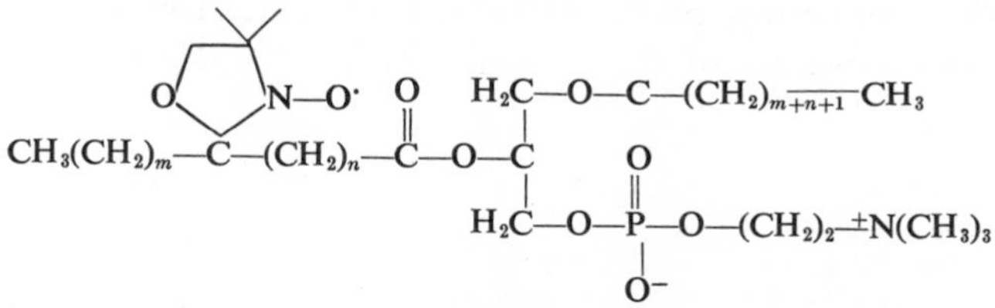

LXIII

Such radicals have an elongated shape (Figs. 68–70) and rotate mainly around their long axis, enhancing the anisotropy of the matrix.

Let us consider the simplest case of rapid rotation of a nitroxide radical around the axis $\bar{R}$, which coincides with the direction of one of the molecular axes (Fig. 71). The values of g_{ik} and A_{ik} characterizing the spectrum are obtained by simple averaging of the respective tensor elements. If, for example, the axis $\bar{R}$ is parallel to the y axis of the radical, the spectrum corresponding to $\bar{H} \parallel y$ remains unchanged and is described by the pa-

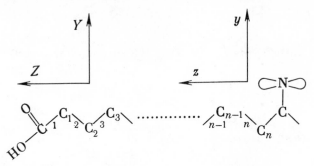

Figure 68. Spin-labeled fatty acid LVII in the stretched conformation. The Z axis is parallel to the polymethylene chain and parallel to the principal z axis of the hyperfine tensor [455].

rameters $A'_\parallel = A_{yy}$ and $g'_\parallel = g_{yy}$. The spectra corresponding to $\bar{H} \parallel x$ and $\bar{H} \parallel z$ become equivalent and are described by the parameters $A'_\perp = \frac{1}{2}(A_{xx} + A_{zz})$ and $g'_\perp = \frac{1}{2}(g_{xx} + g_{zz})$. Thus in those cases in which $\bar{H}$ is parallel to $\bar{R}$, the parameters of the observed spectrum are $A'_\parallel$ and $g'_\parallel$, and when the external field is perpendicular to $\bar{R}$, the parameters of the observed spectrum are $A'_\perp$ and $g'_\perp$. In those cases in which $\bar{R}$ is parallel to x, the spectrum corresponding to $\bar{H} \parallel \bar{R}$ is described by the parameters $A'_\parallel = A_{xx}$ and $g'_\parallel = g_{xx}$, and the spectrum corresponding to $\bar{H} \perp \bar{R}$ is described by the parameters $A'_\perp = \frac{1}{2}(A_{yy} + A_{zz})$ and $g'_\perp = \frac{1}{2}(g_{yy} + g_{zz})$. In those cases in which $\bar{R}$ is parallel to z, we find in a similar manner that $A'_\parallel = A_{zz}$, $A'_\perp = \frac{1}{2}(A_{xx} + A_{yy})$, $g' = g_{zz}$, and $g'_\perp = \frac{1}{2}(g_{xx} + g_{yy})$. The spectrum of a rigid glass is the sum of the three main spectra plus the spectra at all intermediate orientations of the external field $\bar{H}$ relative to the axis $\bar{R}$. Using the known values of the elements of the A and g tensors, we can

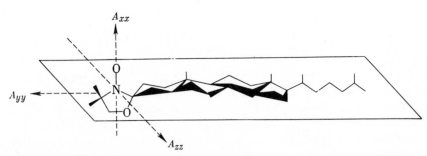

Figure 69. Structural formula of spin probe LX [476]. The probe rotates predominantly about its long axis, which coincides with the principal y axis of the hyperfine tensor.

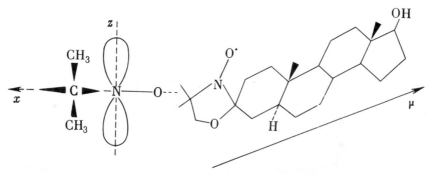

Figure 70. Structure of steroid spin probe LXI [485]. The unpaired electron is localized in the 2pπ orbital of the nitrogen atom. The long axis of the molecule (the main axis of rotation) is perpendicular to the principal z axis of the hyperfine tensor.

calculate the spectrum of the rigid glass and the spectra corresponding to rapid rotation around the x and y or z axes with the aid of a computer [6].

In the general case in which the motion of the radical is characterized by axial symmetry relative to a certain axis $\bar{R}$, its EPR spectrum is described with the aid of the effective spin Hamiltonian

$$\hat{H}' = \beta \hat{S} g' \bar{H} + \hat{S} A' \hat{I} - g_N \beta_N \bar{H} \hat{I}. \qquad (10.1)$$

The elements of the g' and A' tensors are obtained as a result of the appropriate averaging of the elements of the tensors g and A [2,453]. The

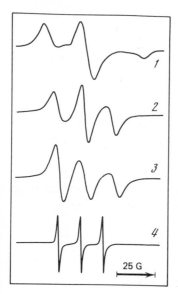

Figure 71. Computer-calculated spectra of nitroxide radicals [6]: (*1*) set of radicals randomly oriented in a rigid matrix; (*2*) rapid anisotropic rotation about the R axis, which is perpendicular to the y axis; (*3*) rapid anisotropic rotation about the R axis, which is parallel to the x axis; (*4*) rapid isotropic rotation.

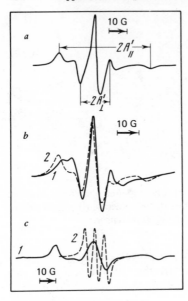

Figure 72. Electron paramagnetic resonance spectra of radicals in various media: (*a*) radical LVII (*m* = 17, *n* = 3) in a sonicated aqueous dispersion of soybean phosphatides, pH 8.0 [2]; (*b*) radical LVII in a walking-leg nerve fiber of *H. americanus* with a parallel (*1*) and a perpendicular (*2*) orientation of the cylindrical axis of the nerve fiber relative to the magnetic field ($\overline{H}$) of the spectrometer [2]; (*c*) radical LXI in a hydrated (16% water) cholesterol–lecithin (1:1) film oriented parallel (*1*) and perpendicular (*2*) to $\overline{H}$ [457].

spectrum of radical LVII (17.3 G) in phosphatide micelles (Fig. 72) corresponds to the case of very high ordering of the radicals in the medium. It is characterized by the quantities $A'_{\parallel} = 26$ G and $A'_{\perp} = 8.5$ G [2]. The similarity between the values of $A'_{\parallel}$ and A_{zz} of this radical (30.7 G) indicates that the angle between the axis $\overline{R}$ and the z axis of the radical is relatively small. The axial axis of rotation of radical LVII is its long axis, which coincides with the z axis of the tensor of hyperfine interaction constants (see Fig. 68). This radical displays rapid anisotropic rotation in nerve-fiber membranes (Fig. 72*b*). Analysis of the spectra obtained with $\overline{H} \parallel \overline{N}$ and $\overline{H} \perp \overline{N}$ ($\overline{N}$ is the vector of the overall cylindrical axis of the axons) reveals that the orientation $\overline{R} \perp \overline{N}$ is preferred almost five to one over the orientation $\overline{R} \parallel \overline{N}$ [2]. This led McConnell and McFarland to the important conclusion that the long axis of the radical is oriented perpendicular to the surface of the membrane and that the oxazolidinyl ring is oriented parallel to the surface.

The degree of ordering in the long axes of nitroxide radicals in a membrane is conveniently characterized with the aid of the semiempirical order parameter S [54,455,456]. If the rotation frequency of a radical around its long axis ν is much greater than 10^8 sec^{-1}, we have

$$S = \frac{A'_{\parallel} - A'_{\perp}}{A_{aa} - A_{bb}} \qquad (10.2)$$

In the case of radicals LVII, LXII, and LXIII, whose long axes coincide with the z axis of the nitroxide fragment, we have

$$A_{aa} = A_{zz} = 30.8 \pm 0.5 \text{ G and } A_{bb} = \tfrac{1}{2}(A_{xx} + A_{yy}) = 5.8 \pm 0.5 \text{ G} \quad (10.3)$$

The long axes in the steroid (LXI) and cholestane (LX) radicals coincide with the y axis of the A and g tensors. For these probes we have

$$A_{aa} = A_{yy} = 5.8 \text{ G}, \qquad A_{bb} = \tfrac{1}{2}(A_{xx} + A_{zz}) = 18.3 \text{ G}. \quad (10.4)$$

In those cases in which it is necessary to take into account the effect of the polarity of the environment of the nitroxide fragment, the right-hand side of Eq. (10.2) is multiplied by the ratio between the isotropic hyperfine interaction constants [54]

$$\frac{A_{iso}}{A'_{iso}} = \frac{\tfrac{1}{3}(A_{xx} + A_{yy} + A_{zz})}{\tfrac{1}{3}(A'_{\parallel} + 2A'_{\perp})} \quad (10.5)$$

If the long axes of the radicals are completely oriented in a membrane, then it follows from Eqs. (10.2)–(10.4) that $S = 1$, while the value of S can deviate significantly from unity in other cases. The more random the motion of radicals and the greater the amplitudes of their motions (the closer the medium is to the liquid state), the closer $A'_{\parallel}$ is to $A'_{\perp}$. In the limiting case of isotropic rotation we have $A'_{\parallel} = A'_{\perp}$ and $S = 0$. Therefore, S is sometimes used as a parameter describing the fluid nature of membranes [54].

Under certain assumptions concerning the nature of the motion of a radical in a medium, the order parameter can be related with averaged values of the angles between the major axes of the radical and a symmetry axis of the medium. For example, if we assume that the long axis of a probe precesses rapidly around a normal to a lipid bilayer membrane (or around an optical axis of a liquid crystal), we have

$$S = \tfrac{1}{2}(3 < \cos^2 \theta > -1), \quad (10.6)$$

where θ is the precession angle [54,456,476]. In the limited random walk model [477,476] the rotation axis of a radical undergoes rapid movements within a cone with an angle of $2\theta'$ (around a normal to the bilayer). In this model we have

$$S = \tfrac{1}{2} (\cos \theta' + \cos^2 \theta'). \quad (10.6a)$$

In experiments with nitroxide derivatives of fatty acids it is necessary to take into account the flexibility of the polymethylene chains, that is, the probability of *trans–gauche* isomeric transitions relative to the C—C bonds. This was successfully done by Hubbell and McConnell [54]. They showed that if the probability of an energetically unfavorable *gauche* conformation (P_g) is small and constant for all the C—C bonds of a chain, then

$$\log S_n = n \log P_c + \log S_0 + C, \tag{10.7}$$

where S_n is the order parameter calculated from Eq. (10.2), n is the number of methylene carbons between the nitroxide fragment and the terminal group of the polymethylene chain, $P_c = (1 - P_g)$ is the probability of the *trans* conformation (that characterizes the degree to which the polymethylene chain can be stretched), S_0 is the order parameter for radicals considered as rigid rods, and C is the iteration constant.

Thus analysis of the EPR spectra gives some idea as to the nature of the motion, orientation, and conformation of nitroxide radicals in membranes.

The determination of $A'_{\|}$ and $A'_{\perp}$, which are needed to calculate S, is a very simple problem, if the hyperfine extrema are well resolved. This is usually the case when the radicals are highly ordered in the membrane (Fig. 72). When the degree of ordering is low (e.g., in investigations of membranes with the aid of probes LIV–LVI), the inner hyperfine extrema merge with the outer, resulting in the appearance of slight asymmetry in the lines of the spectrum. It has been shown [10] that investigation of the slight anisotropy of the EPR lines also makes it possible to determine the order parameter of radicals. In addition, it was shown that in such a case it is possible to find the rotational correlation time of the radical from the usual relationships, that is, to evaluate the local viscosity of an anisotropic system. This approach has been used with success in the study of the phase states of the potassium palmitate–water system [458].

As the polarity of the solvent is increased, there is a small increase in the isotropic hyperfine interaction constant A_{iso}, as well as a decrease in g_0 [38,460]. For example, in the case of di-*tert*-butylnitroxide, the transition from a hydrocarbon (heptane, xylene, etc.) to water is accompanied by an increase in A_{iso} from 14.8 to 16.7 G and a decrease in g_0 from 2.0061 to 2.0056. The spectrum of a probe in a microscopic heterogeneous system such as a biological membrane is usually a superposition of the spectra of radicals located in different parts of the membrane. Sometimes the difference between the polarities of microregions is fairly large, and there is splitting of the spectral components in strong fields (I_{-1}). Figure 73 presents the spectra of radical LIV in pea chloroplast membranes at various pH values. The spectrum with the component I'_{-1} is characterized by a smaller value of A_{iso} and a larger value of g_0 than is the spectrum with the component I''_{-1}. The first of these superimposed spectra is caused by radicals in the more hydrophobic regions of the membrane, and the second is caused by radicals in comparatively less hydrophobic regions of the membrane. Decreasing the pH to 5 or less results in an increase in the contribu-

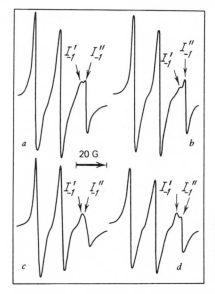

Figure 73. Electron paramagnetic resonance spectra of radical LIV (0.1 mM) in pea-chloroplast membranes (20 mg/ml chlorophyll) at various pH values [461]: (*a*) 8.0; (*b*) 13.0; (*c*) 3.0; (*d*) 5.0.

tion of the component I'_{-1} to the overall spectrum. This indicates that there is an increase in the accessibility of the hydrophobic regions of the membranes to the probe molecules. Alkalinization of the medium causes a redistribution of the probe into less hydrophobic regions of the lipoproteins, which is accompanied by an increase in the contribution of the component I''_{-1} to the overall spectrum [461].

Similar splitting of high-field components is also observed in the spectrum of radical LIX in liver microsomal membranes (Fig. 74). In the general case heterogeneity of a matrix causes broadening of the spectral lines. By comparing such a spectrum with the sum of the spectra calculated with a computer, we can study the distribution of the probe among regions with different polarities and viscosities [459].

Altering the structure of a membrane can result in a redistribution of the probe among regions with different levels of solvation [459,462,463]. This has an effect on the values of g_{ik} and A_{ik} and, in turn, on the calculated values of τ_c and ε. The reason for the observed changes in τ_c and ε can be found by comparing them with the changes in A_{iso} and g_0.

As long as the purpose of the experiment is not a quantitative investigation, any spectral parameter can be used to monitor the changes in the matrix. For example, it is convenient to use the parameter ε and the right-hand side of Eq. (1.5) as effective structure-sensitive parameters, which give some idea as to the extent of the changes.

As the concentration of nitroxide radicals is increased, the EPR lines begin to broaden because of spin-exchange interactions (Fig. 75). Further

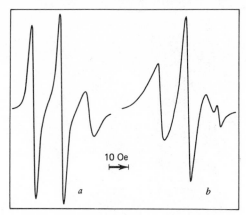

Figure 74. Electron paramagnetic resonance spectra of radical LV (*a*) and radical LIX (*b*) in liver microsomal membranes [462]: radical concentration, 0.1 mM; 20 mg/ml microsomal proteins; 25°; incubation medium, 50 mM Tris-HCl buffer (pH 7.5) and 50 mM KCl [462].

increases in the radical concentration result in exchange narrowing and cause the lines to come closer together. At concentrations on the order of 0.1 M the spectrum becomes a singlet. Undiluted crystals of nitroxide radicals produce singlet spectra. Nitroxide derivatives of fatty acids, steroids, and lipids are practically insoluble in water, and singlet spectra are, therefore, also characteristic for these compounds.

The spectra of radical–aqueous buffer–membrane systems are generally superpositions of a singlet and a triplet signal (Fig. 76). The singlet signal is usually caused by an excess of the radical in the aqueous phase, and in

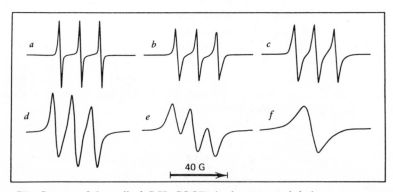

Figure 75. Spectra of the radical $C_9H_{19}COOR_6$ in deoxygenated dodecane at room temperature [5]. Concentration in M: (*a*) 10^{-4}; (*b*) 5×10^{-4}; (*c*) 10^{-3}; (*d*) 3×10^{-2}; (*e*) 6×10^{-2}; (*f*) undiluted solution of the radical.

such a case this signal can be eliminated by washing the membrane free of the radical in an ultracentrifuge [461]. The triplet signal is caused by radicals that are dissolved (solubilized) in the membrane lipoproteins. The empirical "solubilization parameter"

$$\Lambda = L_1/L_0 \qquad (10.8)$$

has been introduced [464] for a quantitative evaluation of the solubilization of a probe by the hydrophobic regions of membranes. Here L_0 is the integral intensity of the central component of the spectrum, which is a superposition of the singlet of the radicals in the aqueous phase and the middle line of the triplet of the solubilized radicals, and L_1 in the integral intensity of a lateral component of the spectrum (both lateral components belong mainly to the triplet of the solubilized radicals and are, therefore, equal in intensity). If the total radical concentration in a series of experiments remains unchanged, the solubilization parameter Λ is unambiguously related to the distribution coefficient of the radical between the lipoproteins and the aqueous phase. In the case of a radical in water $\Lambda = 0$, in a medium in which the radical forms a molecular solution $\Lambda = 1$, and in a suspension of membranes Λ takes on intermediate values [464,465].

Model membrane systems.   Nitroxide radicals were first used [467–468] to study micelle formation in solutions of sodium dodecyl sulfate. In the region of the critical concentration for micelle formation the spectrum of the hydrophobic radical changes from a singlet to a triplet characteristic of radicals solubilized in a hydrocarbon micellar phase. In this region there

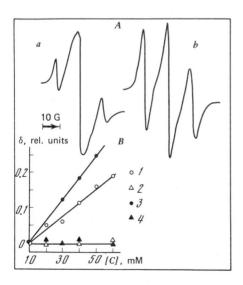

Figure 76.   Electron paramagnetic resonance spectra of radical LV (0.5 mM) in oxidized submitochondrial particles (A) and the relative broadening (δ) of the low-field component of the spectrum of potassium ferricyanide (B): (a,b) in the absence and presence of 10 mM ATP; (1) submitochondrial particles (10 mg/ml) without ATP; (2) same in the presence of 10 mM ATP; (3) HSA (25 mg/ml) without ATP; (4) mitochondrial lipids.

is also a sharp increase in the τ_c value of the radical. The values of the critical concentration for micelle formation determined with the aid of spin probes are in good agreement with the values found by other methods. The values of τ_c for nitroxide radicals in sodium dodecyl sulfate micelles do not exceed 10^{-9} sec, but correspond to very slightly hindered rotation of the probe.

Nitroxide radicals have been used to study structural changes in aqueous solutions of the nonionogenic detergent Twin 80 as a function of its concentration [469]. Interactions between micelles begin to be important in the region of high detergent concentrations. Inversion of the micelles, that is, an oil–water to water–oil structural transition, occurs at a specific concentration of the detergent. Unlike the "normal" micelles, in the inverted micelles the water is located inside of the micelles, and the carbon chains of the detergent molecules are turned outward. The inversion region is readily detected according to characteristic changes in the parameters of the EPR spectrum of a probe. For example, there is a maximum on a plot of τ_c as a function of the detergent concentration in the region of inversion. Analysis of the hyperfine interaction constants of a broad range of radicals has made it possible to establish the dynamic nature of the solubilization of detergents in micelles [467–469].

The rotation of nitroxide radicals in model lipid membranes is characterized by a correlation time on the order of 10^{-9}–10^{-7} sec and considerable anisotropy. The degree of immobilization and anisotropy in the rotation depends on the type of probe, the type of membrane, and the properties of the latter (mobility of the lipids, their packing, etc.).

An aqueous dispersion of egg lecithin and cholesterol (in a 2:1 molar ratio) was studied [54] with the aid of nitroxide derivatives LVII and LXIII with nitroxide fragments on different carbon atoms of the polymethylene chain. The ratios of the molar concentrations of the probes and lipids in similar experiments was varied from 1:150 to 1:500. According to electron photomicrographs, in such amounts nitroxide radicals have practically no effect on the structure of the lipid assemblage [6]. The EPR spectra showed distinct outer extrema with splitting equal to $2A'_{\parallel}$ [54]. This made it possible for Hubbell and McConnell to determine the order parameter of the radical in the liposomes. It turned out that the value of S_{zz} decreases sharply as the nitroxide fragment is shifted from C-4 to C-12. These workers explained the rapid anisotropic rotation of the radical with the aid of a model of two isomeric states of the polymethylene chains undergoing rapid interconversion. The good fit between the theoretical and experimental spectra made it possible to evaluate the probabilities of the *gauche* and *trans* conformers. It was concluded that the portion of the hydrocarbon chain of egg lecithin up to C-8 (counting from the glycerol–fatty acid ester bond) has the structure of a rigid rod, while in the more distant portions of the chains the probability of the *gauche* conformation increases sharply, and the mobility of the ends of the hydrocarbon chains increases in accordance. A similar gradient in the mobility of hydrocarbon chains has been observed in micelles of dipalmitoylphosphatidylcholine (at temperatures above the melting point) and in axonal membranes of *H. americanus* [54]. The existence of a gradient in the mobility of polymethylene chains has also been indicated by the results of an investigation of spin–lattice relaxation of ^{13}C nuclei in lecithin liposomes [470].

Cholesterol decreases the mobility of spin-labeled fatty acids and, to a lesser degree, the mobility of cholestane probes in egg-lecithin liposomes [449,471–473]. In addition, cholesterol increases the anisotropy in the rotation of the radicals. In liposomes of saturated dipalmitoylphosphatidylcholine lipids cholesterol increases the mobility and decreases the ordering of the probes [449,472,473]. The results of the studies on the effect of cholesterol on the mobility of spin probes in liposomes are consistent with the hypothesis that cholesterol decreases the mobility of lipid hydrocarbon chains at temperatures above the melting points of the polymethylene chains but increases it at temperatures below these melting points [474].

The protein rhodopsin, like cholesterol, slows the segmental motion of hydrocarbon chains and increases the order parameter of a phosphatidylcholine spin probe in liposomes and multilayers of phosphatidylcholine [475].

Evaporation of a solution of lecithin (e.g., in chloroform) on a glass or quartz backing produces a film containing layers of spontaneously oriented lipid molecules. Figure 77 shows the EPR spectra of cholestane probe LX in hydrated multilayer films of egg lecithin as a function of the cholesterol concentration in the films and their orientation in the magnetic field of the spectrometer [476]. The hyperfine splitting in the spectra with a perpendicular orientation of the plane of the film relative to H decreases from 9.4 G at 0% cholesterol to 6.7 G (the minimum) at 55% cholesterol and then increases again. The hyperfine splitting for the parallel orientation of the films relative to $\bar{H}$ increases from 17 G (0% cholesterol) to a maximum value of 18.9 G (55%) and then decreases. When the orientation of the film is perpendicular, the hyperfine-structure lines are broadened symmetrically, the magnitude of the broadening being smallest at 55% cholesterol. When the orientation of the film is parallel, the hyperfine structure lines are broadened asymmetrically, and the broadening is greatest at 55% cholesterol. The strong dependence of the spectrum on the orientation of the film in the magnetic field can be explained to a first approximation by assuming that the long axes Y of the radicals are oriented perpendicular to the surface of the film and that the radicals rotate rapidly ($\tau_c \ll 10^{-8}$ sec) around their long Y axes. The rapid rotation around the Y axes averages the x and z components of A and g tensors. The spectrum corresponding to the parallel orientation of the film is then described by the effective tensor components

$$A'_\perp = A'_{zz} = A'_{xx} = \tfrac{1}{2}(A_{zz} + A_{xy}) = 18.3 \text{ G}$$

and

$$g'_\perp = g'_{zz} = g'_{xx} = \tfrac{1}{2}(g_{zz} + g_{xx}) = 2.0055.$$

The spectrum corresponding to the perpendicular orientation of the film is described with the aid of the following parameters:

$$A'_\| = A'_{yy} = A_{yy} = 5.8 \text{ G} \quad \text{and} \quad g'_\| = g'_{yy} = g_{yy} = 2.0058.$$

Lapper, Paterson, and Smith [476] compared the experimentally obtained spectra with spectra calculated on a computer under different assumptions regarding the nature of the motion and the orientation of the probe in the films. A limited random-walk model in which the long axis of the radical randomly precesses and oscillates within a definite angle relative to the normal to the surface of the membranous film appears to

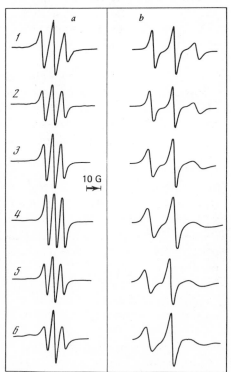

Figure 77. Electron paramagnetic resonance spectra of radical LX in a lecithin film containing various concentrations (0–80%) of cholesterol [476]. The magnetic field (H) is perpendicular (a) and parallel (b) to the plane of the film; the cholesterol concentrations are (in %): (*1*) 0; (*2*) 20; (*3*) 40; (*4*) 50; (*5*) 60; (*6*) 80.

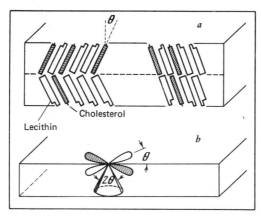

Figure 78. Schematic representation of the orientation of the phospholipids and cholesterol in a bilayer (*a*) and the cone described by the long axis of the probe (*b*) [476]. The
π orbital of the unpaired electron forms the angle θ with the plane of the bilayer.

be the most accurate. On the average, the axis of the radical describes a
figure within a cone with an average angle Θ at its apex (Fig. 78). The
spread in the values of this angle around the mean value (for an ensemble
of radicals) does not exceed 10° and decreases with increasing cholesterol
content in the film. The amplitude of the random deviations of the radical
axis from the optical axis of the film decreases from 46° at 0% to 17° at
55% cholesterol and increases again on further additions of cholesterol.
In addition, as the cholesterol concentration is increased, the rotational
mobility of the probe decreases (τ_c increases from $\sim 10^{-9}$ to 5×10^{-9} sec).

The results indicate that cholesterol decreases the mobility and simultaneously fixes the orientation of the lipids along normals to the film
surface.

Analogous results were obtained [477] in which nitroxide fatty acid
derivatives served as spin probes. It is interesting to note that, as in the
liposomes, the mobility of spin-labeled fatty acids [477] and spin-labeled
lipids [478] in lipid films increases as the nitroxide fragment becomes more
distant from the polar ends of the lipids.

The mobility of the probe in the film increases as the amount of water
increases [457,477]. Hydration of the film has a strong effect in the case of
probes labeled near the polar ends and a lesser effect on the mobility of the
ends of the polymethylene chains [477]. The effect of increasing concentrations of cations on the ratio I_{+1}/I_0 between the amplitudes in the spectrum of a hydrophobic probe in a film of bovine CNS lipids was studied in
[479]. This ratio increased with increasing cation concentration (reaching
a limiting value), indicating, in the opinion of these researchers, an increase

in the ordering of the lipids. The La^{3+} ions were more effective than the Ca^{2+}, and the latter were in turn more effective than the Na^+ ions. Aliphatic alcohols induce changes in the packing of the lipids in a bilayer, the minimum alcohol concentrations that induce significant changes in the EPR spectra being coincident with the concentrations at which these alcohols have an anesthetic effect on biological systems [480].

Mention should be made of the work of Verma, Schneider, and Smith [481], which is one of the first attempts to detect photoinduced structural changes in phospholipid multilayers containing light-sensitive pigments of the chlorophyll type.

In recent years a great deal of attention has been devoted to temperature-induced structural changes in liquid crystals, lipid-water systems, and biological membranes. There has been special interest in phase transitions of the "hydrated crystalline gel–hydrated liquid crystal" type, which result in drastic changes in the physical state and organization of lipid molecules [474]. The transition temperature (T_n) is dependent on the chemical nature of the polar ends of the lipids, the degree of saturation in the hydrocarbon chains, the length of these chains, the amount of water in the system, and additions of cholesterol and other effectors. In the case of systems consisting of dipalmitoylphosphatidylcholine (saturated lipid) and water ($> 30\%$) the value of T_n is 41°. Despite numerous investigations, the physical nature of the transition is still unclear. The data obtained with X-ray diffraction analysis, calorimetry, NMR, and infrared spectroscopy indicate that at temperatures below T_n the hydrocarbon chains exhibit quasicrystalline packing, while at temperatures above T_n they are in a more fluid state [474,482]. There is some information indicating that at temperatures above T_n the polar groups also possess greater rotational mobility [483].

The rotational mobility of nitroxide probes increases sharply in the vicinity of T_n for dipalmitoylphosphatidylcholine [6,54,453,469,477,484–486]. The "melting" of a sonicated aqueous dispersion (liposomes) of this lipid was studied [6,477] with the aid of the nitroxide fatty acid derivatives LVII (m,n). The ratio I_0/I_{+1} served as the structure-sensitive parameter. The parameter of the spectrum of 16-nitroxide stearate LVII (1,14), in which the nitroxide fragment is on the end of the hydrocarbon chain, varies sharply in the vicinity of 41°. In the case of 12-nitroxide stearic acid LVII (5,10) and especially 5-nitroxide stearic acid LVII (12,3), which has the nitroxide fragment at the polar end, the transition region is broader and is shifted toward lower temperatures. These data indicate immobilization in the vicinity of the polar portions of the lipids and greater freedom of motion in the vicinity of the ends of the hydrocarbons. The conclusion has been drawn that increasing the temperature above the melting point causes a significant increase in the mobility of the terminal methyl groups and has a considerably smaller effect on the mobility of the polar portions of the lipids. A quantitative investigation of the temperature dependence of the order parameter of spin-labeled lipids has led to similar conclusions [54].

Long, Hruska, and Geesser [487] studied phase transitions in oriented multilayers (films) of sphingomyelin with the aid of probes LVII and LX. A temperature-induced transition in the 26–38° range with a mean point at 32° was discovered in hydrated films containing cholesterol.

The steroid spin probe LXI was used to study the phase states of monomolecular and bimolecular dipalmitoyllecithin membranes in water [485, 486]. In these membranes radical LXI has its long axis oriented along the hydrocarbon chains (Fig. 79). The rotation of the probe is almost isotropic, that is, the distance between the outermost lines of the spectrum (32 G) practically coincides with twice the value of the isotropic hyperfine interaction constant ($2A_{iso} = 28.2$ G). The rotational correlation time at 20° is about 10^{-8} sec, and at temperatures above T_n the value of τ_c decreases by a factor of 3–4. In the region of the crystal–liquid crystal transition (30–40°) there is sharp narrowing of the EPR lines of the probe, indicating an increase in the mobility of the hydrocarbon chains in the membrane at temperatures above T_n.

Besides the hydrophobic steroid spin probe, which binds with the hydrocarbon portions of membranes, the optical probes 8-anilino-1-naphthalene sulfonate (ANS) and bromthymol blue (BTB), which bind with the polar portions of lipids, have been used [485,486]. The "melting" of the membranes is accompanied by a sharp increase in the intensity of the fluorescence of ANS and a decrease in the optical absorption of BTB. The changes in the optical characteristics of the dyes are due to the sharp increase re-

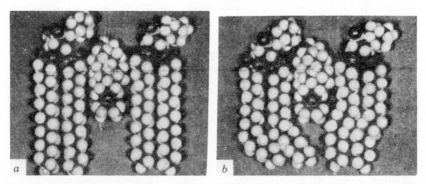

Figure 79. Pauling–Corey–Koltum model depicting a molecule of steroid spin probe LXI between two lipid molecules [485]: (*a*) below the melting point of the lipids; (*b*) above the melting point.

sulting from the phase transition in the number of sections of the membrane surfaces that are capable of binding the ANS and BTB molecules. These data indicate that the polar groups of the lipids are packed considerably more densely at temperatures above T_n than at temperatures below T_n.

In the experiments with the spin probe, various radical concentrations, ranging from 0.035 (probe-to-ligand molar ratio) to 0.27, were employed. Accordingly, the mean distances between the radicals varied as a function of their concentration from 40.6 to 16.1 Å. At higher radical concentrations and temperatures below T_n the lines of the spectrum are broadened and practically merge into a singlet (Fig. 80). However, at temperatures above T_n the EPR spectrum of the probe is a well resolved triplet, even at high concentrations. Comparing the experimental data with the theoretical, Sackmann and Träuble showed that the main reason for broadening of the lines is the exchange interaction between the unpaired electrons and determined the exchange frequency W_{ex} at different temperatures and probe concentrations. The values of W_{ex} were equal to 10^6 to 10^7 Hz. The phase transition in the vicinity of T_n caused W_{ex} to change by 80%. Moreover, there were completely different functional relationships between W_{ex} and the probe concentration at temperatures below and above the transition point. In the opinion of these workers, at temperatures above T_n, that is, in the liquid–crystal state of the membrane, the molecules of the lipid and the steroid probe act as an ideal mixture. The probe molecules (and probably the lipid molecules) are capable of translational (lateral) diffusion in the plane of the membrane. The diffusion coefficient calculated from the value of the pairwise exchange frequency is 10^{-8} cm^2 sec^{-1}. This means that a molecule undergoing diffusion in the plane of the membrane travels a distance of about 20,000 Å sec^{-1}. Below T_n translational diffusion is impossible, although the rotational mobility of the molecules is still very high ($\nu_c = 1/\tau_c \sim 10^8$ sec^{-1}). The molecular organization of the mixed system at temperatures below T_n can be described as a mosaic structure, in which small clusters of steroid molecules are statistically included in a lipid matrix. The concentration of the clusters, their shape, their size, and the number of steroid molecules in a cluster can be determined from the EPR spectra of the probe. The density of the clusters is not dependent on the steroid-to-lipid molar ratio. At 19° it equals 3.35×10^{11} cm^{-2}. The clusters grow with increasing steroid concentration. At $[C] = 0.035$ M the radius of the clusters is 30 Å and contains 65 steroid molecules, and the average distance between the clusters is 285 Å. At temperatures above T_n the clusters dissolve, and the membrane becomes homogeneous. The phase transition is reversible. The transition results in expansion of the lattice. The magnitude of this expansion and the temperature dependence of the lattice constant can also be determined from the temperature dependences of the spectra of the spin probe.

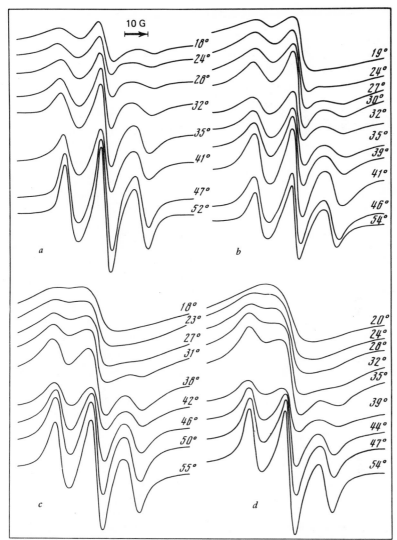

Figure 80. Electron paramagnetic resonance spectra of probe LXI in the lipid phase of multilayer dipalmitoyllecithin at various temperatures and probe concentrations [485]: (a) molar probe-to-lipid ratio (C) equal to 0.035; average distance between probe molecules (d) equal to 40.6 Å; (b) C = 0.075, d = 28.2 Å; (c) C = 0.13; d = 22.0 Å; (d) C = 0.27, d = 16.0 Å.

A different experimental technique was used [488] to determine the coefficient of translational diffusion of spin-labeled phosphatidylcholine (LXII) in oriented multilayers of dihydrostearylphosphatidylcholine and lecithin. Small (about 1 mm diameter) drops of radical LXII were impregnated into films of dihydrostearylphosphatidylcholine and lecithin. As the radicals diffused parallel to the plane of the films, the spectrum changed from a singlet characteristic of large local radical concentrations to a triplet. The time needed to reach equilibrium varied from 10–20 hr to 2–3 days, depending on method of preparing the film and the temperature. The coefficient of translational diffusion is equal to approximately 10^{-8} cm^2 sec^{-1} at 25°, that is, on the same order as the coefficient of translational diffusion of the spin-labeled steroid in [485,486].

The distribution of the spin-labeled phosphatidylcholine (LXII) in bilayer membranes of egg lecithin was studied [489] by "titrating" the EPR signal with sodium ascorbate. The addition of ascorbate decreased the EPR signal by 65%, and it was thus concluded that 65% of the total number of probe molecules are located in the outer membrane layers accessible to ascorbate and that 35% are in the inner layers into which the hydrophilic ascorbate molecules do not penetrate. The asymmetry of the distribution of paramagnetic molecules between monolayers decreased over the course of several hours. A thorough investigation of the changes in the intensities of the signals and in the shape of the spectrum led to the conclusion that there is exchange by the molecules of the spin-labeled lipid between the inner and outer layers of the membranes. At 30° the probability of a flip-flop transition from an inner layer to an outer is 0.07 hr^{-1}, and the probability of the reverse transition is 0.04 hr^{-1}.

Protein–lipid complexes. The formation of a complex between phospholipids and cytochrome C with a maleimide spin label on the ε-amino lysine groups has been studied [490]. The formation of the complex was accompanied by considerable immobilization of the label. After the complex was transferred to isooctane, the rotation of the label was slowed still further. An increase in the correlation time of spin-labeled fatty acids in lysolecithin dispersions following the addition of the erythrocyte apoprotein serum albumin or casein has been noted in many papers [484,491–493].

The effect of lipids on the mobility of blood-serum lipoproteins modified by a maleimide label has been studied [494]. The addition of the label to the protein did not alter the solubility, the circular dichroism, or the immunochemical properties of the preparation. The spectrum of the lipoprotein had two components, one from the strongly immobilized labels and one from the slightly immobilized. Removal of the lipid resulted in appreciable weakening of the strongly immobilized component, this change being irreversible. It was concluded that the existence of this component is due to a specific protein–lipid interaction, which is irreversibly lost as a result of the extraction of the phospholipids from the lipoprotein.

As a whole, the considerable immobilization of spin-labeled fatty acids, steroids, and phosphatidylcholine in complexes of lysolecithin with casein [493], blood-plasma lipoproteins [491,494,495], and in protein–lipid complexes of erythrocyte [493] and *Mycoplasma laidlawii* [496] membranes indicates that the mobility of the hydrocarbon chains of the lipids in them is very low.

The interaction between ferricytochrome C with an iodoacetamide label covalently bound to the methionine-65 residue and mitochondrial membranes has been studied [524]. The binding of the cytochrome C by the mitochondria was accompanied by an appreciable increase in the τ_c value of the label, from 9.3×10^{-10} to 3.3×10^{-9} sec. The isotropic hyperfine interaction constant, however, remained unchanged and equal to 16.7 G, and it was thus concluded that both in the free and in the membrane-bound protein the label is in a moderately polar environment. An increase in the viscosity of the medium (by adding sucrose) was accompanied by an increase in the τ_c value of the label in the case of the free cytochrome, but in the case of the membrane-bound cytochrome the value of τ_c was independent of the viscosity of the medium. It appears that the portions of the membrane that bind the cytochrome C are shielded from the external medium. It is interesting that native and spin-labeled cytochrome C, when added to mitochondria devoid of endogeneous cytochrome C restored their respiration to the same extent.

Model membranous complexes consisting of ferricytochrome C, water, and phosphatidylethanolamine or phosphatidylcholine were studied [497] with the aid of radicals LII and LIII. Radical LII became localized in the polar portions of the membranes (the isotropic hyperfine interaction constant A_{iso} was 16.9 G), and radical LIII was in the hydrophobic portions of the membranes ($A_{iso} = 15.3$ G). The behavior of the radicals was very different. For example, radical LIII remained stable over the course of several months at room temperature, while radical LII was reduced within 1–2 days. As the temperature was increased in the 0–45° range, the correlation time of radical LIII decreased exponentially with an activation energy of 3.3–3.4 kcal/mole, while the correlation time of radical LII increased with the temperature. In the -5 to $-10°$ temperature range the decrease in T was accompanied by a sharp increase in the correlation time of radical LII with a formal activation energy of about 40 kcal/mole. In this same temperature range there was a sharp decrease in the dielectric losses in the sample. Nevertheless, the EPR spectra of radical LIII changed only slightly in this temperature range. It was concluded that a phase transition involving reversible solidification of the polar membrane phase formed from water, the polar tips of the lipids, and the polar portions of the protein occurs in the -5 to $-10°$ temperature range. This same paper also presents interesting results of a study of protein–water and lipid–water binary systems at various component concentrations.

Thus the use of spin labels and probes yields valuable information on the stoichiometry and structure of protein–lipid complexes, as well as on the local conformational changes and the specifics of the protein–lipid interactions in them.

Membranes of organelles. Spin probes are most often introduced into membranes by means of diffusion from aqueous, water–alcohol, or alcoholic solutions. In two studies [498,499] a hydrophobic probe was introduced by means of exchange between membranes and spin-labeled serum albumin. Spin labels and probes can be introduced into membranes by biosynthetic pathways. For example, *Neurospora crassa* synthesizes spin-labeled phospholipids when it is grown on a medium containing 12-doxyl-stearate [451]. Yeast cells of *Saccharomyces cerevisiae* incorporate this same label mainly in the neutral lipids of their membranes [500]. Bacterial cells of *Mycoplasma laidlawii* readily esterify exogeneous fatty acids, including spin-labeled fatty acids, to phospholipids and glycolipids [496,501]. Liver microsomal enzymes catalyze the conversion of spin-labeled stearate into spin-labeled phosphatidic acid [502]. Spin-labeled lipids synthesized by cells have been extracted, purified, and used for further studies as spin probes [502,503].

The rotation of nitroxide derivatives of fatty acids is characterized by τ_c values in the 10^{-9}–10^{-8} sec range and by considerable anisotropy. The mobility of the 8-nitroxide stearic acid is higher than the mobility of the 5-nitroxide stearic acid, but lower than the mobility of 12-nitroxide stearic acid [496]. Thus, as in the case of the model lipid membranes, the mobility of the polymethylene chains of the lipids increases with increasing distance from the polar portions of the lipids.

McConnell and his co-workers [498] investigated the EPR spectra of 3-nitroxide 5α-androstan-17β-ol (LX) and 12-nitroxide stearic acid [LVII (5,10)] in erythrocyte membranes oriented in a hydrodynamic field. The hyperfine structure of the spectra was substantially dependent on the orientation of the current relative to the external magnetic field. Investigation of the orientational dependence of the hyperfine structure showed that the long axes of the radicals are directed perpendicular to the lenticular surface of the cell. After studying the orientational dependence of the hyperfine structure or the spectra of nitroxide radicals in the leg axon of the lobster *Homarus americanus*, these workers found that the long axes of the radicals in these membranes are oriented perpendicular to the cylindrical surface [2]. The similarity between the spectra of hydrophobic probes in artificial and intact membranes and the results of the investigations of the dependence of spectra on the orientation of the membranes in the external magnetic field should be considered as proof of the existence of lipid bilayer structures in erythrocyte and axonal membranes.

Analogous data led to the conclusion that there are lipid bilayer structures in several oncogenic and nononcogenic viruses [499].

The rotational mobility of a probe is greatly dependent on the amount of water in the membrane. During the hydration of lyophilized chloroplast and mitochondrial membranes, as well as plasma membranes of the blue–green algae *Nostoc*, the plot of τ_c for probe LII as a function of the water content passes through a maximum when there is 10–20% water in the membrane [48]. When the water content in the membrane is greater than 20%, a monotonic decrease in τ_c is observed. The maximum in τ_c indicates that there is a structural change in the membrane, resulting in an increase in the number of hydrophilic groups in contact with the water and an increase in the sorption surface. The decrease in the free volume needed for each probe molecule results in a decrease in the mobility of the probe and an increase in τ_c. Comparison of these results with the results of investigations of model systems [469,484,538], as well as data from X-ray diffraction analysis [504] and NMR [505] makes it possible to relate the structural change at 10% water to the formation of a layer of structured water and the second change to a change in the phase state of the lipid regions of the membranes.

When the water concentration in the system is above 10%, reduction of the nitroxide radicals by endogenous electron donors begins [48]. The structural transformation when there is 20% water in the membrane is also reflected in the reduction kinetics of the radicals. In these experiments the nitroxide radicals acted both as structural probes and redox electron-acceptor probes.

The effect of γ-radiation on the sorption of water by chloroplast membranes was discovered by Kutlakhmedov et al. [506]. After irradiation of the chloroplasts with doses of up to 300 krad, there was an increase in the correlation time of probe LII. Irradiation with doses greater than 500 krad caused serious denaturation changes in the membranes, as a result of which the maxima on the kinetic curves for the sorption of water disappeared.

In one study [507] nitroxide stearates were used to study the interaction between local anesthetics and the unmyelinated nerve fiber of *H. americanus*. Among the anesthetics tested (azerin, carbocaine, lidocaine, prylocaine, nupercaine, and procaine), only the acetylcholinesterase inhibitor procaine had an effect on the EPR spectra. The increase in the mobility of the probe under its effect was attributed to procaine-induced conformational changes in the membrane involving the entire region of the hydrocarbon chains of the lipids.

A decrease in the local viscosity of the hydrophobic portions and an increase in the solubilization of nitroxide probes in the erythrocyte stroma under the effect of benzyl alcohol, narcotics (tetracaine, etc.), higher alcohols, and butyl mercaptan were noted in another study [508]. Potassium, calcium, and other ions decreased the distribution coefficient of radicals with a cationic group near the nitroxide fragment. A radical with a quaternary ammonium group was more sensitive to hydrophilic additions, and a probe with a nitroxide fragment in the hydrocarbon chain was more sensitive to hydrophobic additions.

According to certain data [509], increasing the amount of cholesterol in an erythrocyte suspension causes contraction of the lipid portions of the membranes. The local viscosity of the lipid portions of liver endoplasmic reticulum membranes decreases under the effect of Ca^{2+}. The curve describing this decrease has a two-phase nature. In low concentrations the Ca^{2+} ions interact with the lipoprotein subunits of the membranes, and in high concentrations they interact with the lipids [461].

Nitroxide radicals LIV–LVI and LIX have been used to detect and study conformational changes in mitochondrial membranes under the effect of ATP [461,464–466]. Mitochondrial preparations contain many endogenous electron donors that reduce nitroxide radicals. Therefore, potassium ferricyanide was added to the preparations to oxidize the endogenous reducing agents and regenerate the probes. According to Fig. 76, the addition of ATP to mitochondrial preparations causes an increase in the contribution of the triplet to the total spectrum of the hydrophobic probe. This indicates an increase in the solubilization of the probe in these membranes under the effect of ATP. Since under the conditions of these experiments the probe molecules were electrically neutral, the effect of ATP can hardly be attributed to a redistribution of the probe between the outer and inner volumes of the membranes due to changes in ionic transport in the presence of ATP. The improved solubility of the probe in the membranes is due to changes in the structure of the membranes resulting from a conformational change initiated by the hydrolysis of ATP on mitochondrial ATPase. The empirical solubilization parameter Λ (see p. 199) was used to obtain a semiquantitative evaluation of the solubilization of the probe by the membranes. It was found that the dependence of the solubilization parameter on the ATP concentration is S-shaped with a Hill coefficient equal to 6 for the middle part of the curve. The inhibition of the conformational change by dinitrophenol is characterized by the same value of the cooperativity parameter. Oligomycin in small concentrations activates the effect of ATP and decreases the cooperativity parameter like an allosteric effector.

Study of the dependence of the solubilization of a probe on the ambient pH showed that decreasing the pH to 4–4.5 causes a considerable increase in the solubilization of the probe in the membranes without any additions of ATP (Fig. 81). Since the submitochondrial particles (SMP) accumulate protons on addition of ATP [510], it could have been assumed that ATP acts on the membrane conformation indirectly by acidifying the inner portion of the particles. It was, therefore, essential to compare the effect of ATP on the SMP and intact mitochondria, which are characterized by an orientation of the inner membrane opposite that of the SMP and expel protons into the external medium rather than accumulate them. In two studies [464–465] it was shown that ATP increases the solubilization of a probe in both SMP and in intact mitochondria. This effect cannot be due to the accumulation of protons in the external medium, since incubation of the mitochondria with ATP caused practically no change in the pH of the external medium. Moreover, ADP, unlike ATP, did not induce any noticeable changes in the EPR spectra. These experiments imply that the conformational changes observed with the aid of spin probes in membrane lipoproteins are a direct result of the energizing of the membranes by ATP.

Figure 76 illustrates the effect of potassium ferricyanide on the width of the I_{+1} spectral component. In this concentration range ferricyanide can broaden the spectral component without altering its integral intensity in

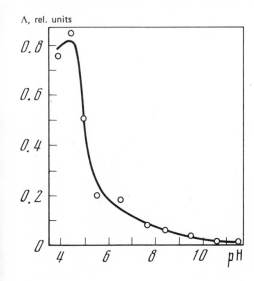

Figure 81. Variation in the solubilization parameter (Λ) of probe LV (2 $\times$ 10^{-4} M) in submitochondrial particles (10 mg/ml protein) as a function of the pH [461].

accordance with Eq. (6.1). Ferricyanide effectively broadens the I_{+1} spectral component in nonenergized SMP, but does not broaden it in energized SMP. This means that in the nonenergized particles the probe is solubilized by the external portions of the membranes that are accessible to ferricyanide and that the addition of ATP causes the probe to move into the inner portions that are not accessible to ferricyanide. Since ferricyanide markedly broadens the spectral components of a radical bound by albumin but does not broaden its spectrum in mitochondrial lipids, it may be assumed that the radicals are solubilized by the outer proteins in the nonenergized particles and by the membrane lipids in the energized particles [461].

In some of the experiments no potassium ferricyanide was added to the SMP. After ATP was added to SMP containing an endogenous oxidation substrate, there was a decrease in the integral intensity of the EPR spectrum. This decrease is clearly due to an increase in the accessibility of the electron donors of the outer portions of the SMP to the probe molecules [461,466].

We note that a recently published paper [511] reported an ATP-initiated increase in the solubilization of spin-labeled lipids in mitochondrial membranes.

There is some increase in the solubilization of hydrophobic probes as mitochondrial membranes age [461]. In this case the triplet part of the spectrum is easily broadened by potassium ferricyanide (Fig. 82).

Spin probes LV and LVI have been successfully used to monitor the conformational changes in liver microsomal membranes during changes in the redox state of the microsomes. A dependence of the solubilization of the probes on the redox state of the microsomal electron-transfer chain was observed [463,512].

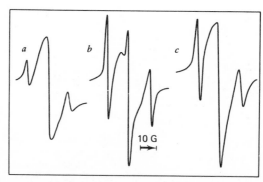

Figure 82. Electron paramagnetic resonance spectra of probe LV [461]: (*a*) in intact submitochondrial particles; (*b*) after 10 hr of incubation at 25°; (*c*) after 10 hr of incubation at 25° and addition of 20 mM potassium ferricyanide.

McConnell et al. [513] found an empirical relationship between the solubilization of radical LIII in membranes and the intensity of the molecular motion of the hydrocarbon chains of the lipids as evaluated from the spectra of nitroxide stearate LVII (5,10), which has a nitroxide fragment in the middle of its hydrocarbon chain. These workers suggested using this relationship to determine the relative concentration of "fluid" lipids, that is, lipids with mobile hydrocarbon chains, in membranes. The concentration of "fluid" lipids in sarcoplasmic reticulum membranes evaluated in this manner is 84% (at 25°). However, additional investigations, particularly an evaluation of the amount of "fluid" lipids in the membrane by some other method, are needed to test the validity of the assumptions underlying this method.

A number of studies have been made of membrane proteins with the aid of covalently bound spin labels. Erythrocyte membranes modified with a maleimide spin label have been investigated [514]. The label reacted specifically with the SH groups of the proteins according to experiments on the blocking of these groups with N-ethylmaleimide. The EPR spectrum displayed signals of both strongly and slightly immobilized labels. These workers therefore concluded that there are two types of SH groups in the membranes. The small reversible changes in the slightly immobilized component under the effect of chlorpromazine were attributed to conformational changes in the membranes under the effect of chlorpromazine.

Changes in the spectrum of erythrocyte ghosts modified with a maleimide label as a function of the pH were observed in one study [515]. A plot of the amplitude of the narrow component of the low-field line of the spectrum as a function of the pH had a minimum in the vicinity of the isoelectric point of the membrane proteins.

Johnes et al. [516] reported that there are small reversible changes in the spectra of ethyl anhydride and maleimide labels, added to the membranes of the air vacuoles of blue–green algae, under the effect of changes in the hydrostatic pressure, temperature, and pH, as well as under the action of trypsin, guanidine chloride, and chloroform.

The interaction between ATP and glycerinated muscle fibers modified by maleimide, isothiocyanate, or iodoacetamide labels has been studied. The ATP-initiated contraction of the muscle fibers was accompanied by a decrease in the mobility of the maleimide label [517].

Spin-labeled analogs of the organic fluorophosphates, which are specific inhibitors of acetylcholinesterases, were synthesized [518] to study acetylcholinesterases bound to erythrocyte and neuronal receptor membranes. These labels completely inhibited the cholinesterase activity of the membranes. The EPR spectra indicated that the rotation of the labels is slightly hindered. The spectra of the same labels added to the active centers of α-chymotrypsin displayed strongly hindered rotation. These workers concluded that the active center of the membrane cholinesterases, unlike that of α-chymotrypsin, is on the surface of the enzyme. This conclusion is, of course, only preliminary. It could be confirmed by "titrating" the spectra of the labels with potassium ferricyanide or ascorbic acid.

The EPR spectra of iodoacetamide and isocyanate labels covalently bound to proteins of sarcoplasmic reticulum membranes were studied in two investigations [519,520]. The spin-labeled preparations maintained their ATPase activity and their ATP-dependent absorption of Ca^{2+}. An addition of ATP caused some small reversible changes in the spectrum of the iodoacetamide label, which were attributed to a conformational change in the vicinity of the labeled SH groups in the membrane proteins. Changes were observed in the spectra of the labels under the effect of ADP, UTP, urea, deoxycholic acid, pH changes, and temperature variations. The maleimide spin label, like the iodoacetamide label, also makes it possible to monitor conformational changes in the sarcoplasmic reticulum under the effect of ATP in the presence of Mg^{2+} and Ca^{2+} ions [521].

Nitroxide derivatives of PCMB have been used to study the role of the SH groups of proteins during the activation of bacterial spores of *Thermoactinomyces vulgaris* [522,523]. The intensity of the signal of the labels bound to the activated spores was almost an order of magnitude greater than the intensity of the signal in the case of the quiescent spores. Thus the activation was accompanied by a considerable increase in the number of SH groups capable of reacting with the labels. Titration of the EPR spectra with potassium ferricyanide showed that the proteins in the membranes of these spores contain at least three classes of topographically dissimilar SH groups with different EPR spectra (Fig. 83). These groups differ with respect to their abilities to react with labels, the mobilities of labels added to them, and their accessibility to ferricyanide molecules and endogenous reducing agents.

The temperature dependences of the parameters of the EPR spectra of hydrophobic probes in a number of biomembranes have revealed temperature-dependent structural changes in the bacterial membranes of *Micrococcus lysodeikticus* and *Escherichia coli* [525,526], *Mycoplasma laidlawii* [496,501], and *Actinomyces streptomycini* B-6 [527], as well as in the electron-transferring membranes of liver microsomes [463,512,528,529], chloroplasts [461,528,529], and mitochondria [461,528–531]. Figure 84 depicts the temperature dependence of the solubilization parameter of probe LV in mitochondrial membranes. Below the transition point (17°) an increase in temperature causes a decrease in the solubilization parameter. Above the transition point an increase in temperature is accompanied by an increase in the solubilization of the probe in the membrane. It appears that

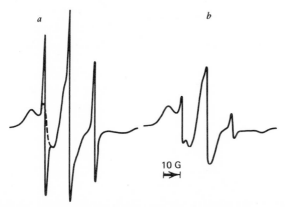

Figure 83. Electron paramagnetic resonance spectra of activated *T. vulgaris* spores co-valently labeled with nitroxide derivatives of PCMB [523]: (*a*) without potassium ferri-cyanide; (*b*) in the presence of 3 mM potassium ferricyanide, 25°, pH 7.5.

at 17° the membrane undergoes a structural transformation, resulting in an increase in the accessibility of the hydrophobic portions of the membrane to hydrophobic probe molecules. Moreover, according to the data in one study [532], the formal activation energies of several reactions that are catalyzed by mitochondrial enzymes (the oxidation of succinate, oxidative phosphorylation, ADP–ATP exchange, and ATP hydrolysis) have a dis-continuous decrease at temperatures of 16–18°.

The transition temperature depends on the functional state of the mem-branes. For example, quiescent spores of *Actinomyces streptomycini* B-6 display a temperature-dependent transition at 55°, while active spores dis-play two transitions, one in the 25–30° range and the other in the 50–60° range. It is important to note that the range of the first transition is the range for optimal growth of the culture and that there is destruction of the spores in aqueous media in the 50–60° temperature range [527]. Further-more, probe LIX, which binds to membrane proteins, is far more sensitive to temperature-induced transitions than are probes LIV–LVI, which bind to lipids [525,526,528,461]. Probe LIV, in turn, with the shortest poly-methylene chain, is more sensitive to these transitions than are probes LV and LVI with their longer polymethylene chains. The temperature-induced transitions in membranes may most strongly affect the regions of contact between the proteins and the lipids. The data obtained by calorimetric [533], X-ray diffraction [534], radiospectroscopic [496,501], and physio-logical [535] investigations indicate an intimate relationship between tem-perature-induced transitions and the chemical nature of membrane lipids. However, the physical nature of temperature-dependent transitions in bio-logical membranes is still unclear.

Λ, rel. units

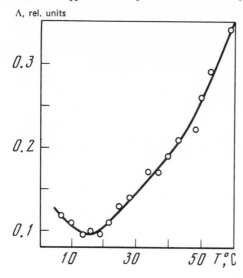

Figure 84. Variation in the solubilization parameter (Λ) of probe LV (2.7 × 10^{-4} M) in submitochondrial particles (10 mg/ml) as a function of the temperature [529].

In conclusion, we shall consider the results of one study [536] in which the spin-exchange constants of radicals LVII (12,3) and LXI in kidney-plasma membranes were measured. The exchange constants, determined as twice the proportionality coefficient between the exchange broadening of the spectrum and the radical concentration, were approximately the same for both probes and equal to 10^7 M^{-1} sec^{-1} at room temperature. This value is of the same order of magnitude as the pairwise exchange constant in model lipid membranes [447,448,488]. If we recall that the rate constant for diffusion-limited reactions in aqueous solutions is 10^9–10^{10} M^{-1} sec^{-1} [537] and assume that this constant is inversely proportional to the viscosity of the solution, it turns out that the viscosity of the biological membrane is 100–1000 times greater than the viscosity of water. The comparatively high values for the exchange constants and the low values for the correlation times of hydrophobic probes are consistent with the theories on the fluid nature of the hydrocarbon portions of biological membranes.

Reduction of nitroxide radicals in cellular preparations.   Nitroxide radicals are effective electron acceptors. Protons are needed to complete the reduction reactions [36]. In media containing protons the role of the reducing agent can be played by any compound with a fairly high reduction potential, such as formic acid, ascorbic acid, hydrazine, hydroquinone, and compounds with SH groups. Nitroxide radicals are reduced to hard to accommodate N-hydroxylamines. The latter can readily be oxidized to the original radicals by oxygen in the air or potassium ferricyanide. The reduction of nitroxide radicals in several inorganic media and lyotropic liquid crystals has been studied [539].

The reduction of nitroxide radicals in biological membranes and organelles was observed even in the first investigations with spin probes [48,51,451,540,541]. In bacterial chloroplasts and chromatophores the kinetics of the photoreduction of nitroxide radicals at low concentrations is described by a first-order equation with respect to the radical concentration [542]. The longer the hydrocarbon chain [461], the slower is the rate of radical reduction. This appears to be due to the lower mobility of radicals with long hydrocarbon chains. A symbatic relationship has been observed between the photoreduction rate of the radicals and the intensity of the EPR signal of the pigment P-890 in the chromatophores [542] and of the pigment P-700 in the chloroplasts [461]. For example, inactivation of the chromatophores under the effect of a detergent or the temperature is accompanied by symbatic changes in the rate constants for the photoreduction of nitroxide radicals, the EPR signal of pigment P-890 (Fig. 85), and the photoinduced changes in the optical density at 890 nm [542]. It may be concluded that the photoreduction of nitroxide radicals is closely related to the functioning of the photochemical centers of the chloroplasts and chromatophores.

Direct manometric experiments have revealed that radical LII is a reactant for the Hill reaction, which is only slightly less active than potassium ferricyanide [461]. Koshland et al. [543,544] demonstrated the destruction of nitroxide radicals in photoirradiated aqueous solutions of serum albumin, tryptophan, tyrosine, and cysteine apparently due to the transfer of electrons from the photoexcited aromatic amino acids and cysteine to the nitroxide fragments. These workers suggested using nitroxide radicals to detect short-lived paramagnetic states in proteins. The reduction of nitroxide radical LII in chloroplasts (but not in chloroplast fragments) and mitochondria was observed in one study [544]. In the mitochondria the reduction was accelerated following the addition of succinate and ADP.

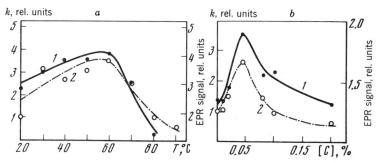

Figure 85. Variation in the photoinduced EPR signal of P-890 (*1*) and rates of the photoreduction (*k*) (*2*) of radical LII by Chromatium chromatophores as a function of the thermoinactivation temperature (*a*) and of the concentration of Triton X-100 (*b*) [542].

According to the results in another study [545], in mitochondrial preparations nitroxide radicals bypass the electron-transfer chain between the points of action of rotenone and antimycin A. The direct electron donors may be the SH groups in the proteins or the hydroquinone compounds in the cells [178,461].

The ability of nitroxide radicals to bypass the electron-transfer chain explains the apparent disruption of photosynthetic phosphorylation by hydrophobic radicals observed by Tsapin et al. [546].

The following can be stated regarding the mechanism for the reduction of radicals during the hydration of lyophilized membranes. When lyophilized preparations are brought into contact with the air, a free-radical EPR signal with $g = 2.0057$ and $\Delta H = 8$ G appears. This signal is caused by semiquinone radicals that arise with the reduction of quinone compounds or the oxidation of cellular hydroquinones [547–549]. The symbatic relationship between the kinetics of the change in this EPR signal and the kinetics of the reduction of the nitroxide radicals during the hydration of organelles can be pointed out [461,547]. The radicals can be reduced directly by the hydroquinones, which are oxidized to semiquinones, but it is possible that the radicals and the quinones are reduced by the same electron donors, such as protein SH groups. Since protons are needed for the reduction of nitroxide radicals, the sharp increase in the reduction rate resulting from the structural change occurring when the water content in the membrane reaches 10% is probably due to an increase in the mobility of the protons in the layer of structured water on the surfaces of biopolymers by a Grotgus mechanism [550]. The accelerated reduction at water contents of 20% or more is most likely due to an increase in the mobility of the radicals as a result of rearrangements in the lipid portions of the membrane [461].

The functional value of the pools of quinone compounds that reduce the nitroxide radicals may be that they are part of the buffer systems that maintain a specific physiological redox state in cells by removing the excess oxidizing agents (endogenous or exogenous) [461]. In this case the undesirable oxidizing agent is a nitroxide radical. These compounds react with oxygen that penetrates into the organelles through damaged (e.g., by lyophilization) membranes and produce an EPR signal with a half-width of 8 G and $g \sim 2$. Other buffer systems, such as the polyphenol oxidase systems of the cell, may oxidize the excess reducing agents. Information on these oxidation–reduction systems, in other words, the participants in the negative feedback in the oxidation-reduction processes in the cell, may be provided by kinetic investigations of the EPR signals as the cells pass from one state to another.

The reduction of nitroxide radicals in biological membranes makes it more difficult to explain the results of experiments with spin probes. Usually the spectrum of a probe in a membrane is a superposition of the spectra of radicals located in portions of different local viscosity and polarity. The reduction of the radicals in the different portions proceeds with different rates. The resultant changes in the spectrum may be caused not only by conformational changes following the addition of an effector, but more specifically by the reduction of the radicals. It is, therefore, necessary to create experimental conditions under which the reduction of the radicals proceeds more slowly than the conformational changes. Sometimes it is possible to prevent the reduction of the probe by adding a particular oxidizing agent to the preparation [461,464].

A very large number of papers have dealt with the study of the effect of nitroxide radicals of various structure on biological and physiological processes in cells and organs of plants and animals. For example, 2,2,6,6-tetramethyl-4-hydroxypiperidine-1-oxyl decreases the number of cells with chromosomal aberrations under γ-irradiation in argon [551]. Di-*tert*-butyl-nitroxide and 2,2,6,6-tetramethylpiperidine-1-oxyl increase the sensitivity of *E. coli* cells to X-irradiation under anaerobic conditions [552]. Radiosensitization of anoxic *E. coli* cells with nitroxide radicals was studied in detail in one investigation [553]. The antineoplastic activity of several nitroxide radicals has been discovered [554]. However, this work is beyond the scope of a review of the application of nitroxide radicals to the molecular biology of membranes.

Thus nitroxide probes yield valuable information of the molecular organization and mobility of biological membranes. The prospective applications of nitroxide radicals in the study of intact biological membranes are greatly dependent on efforts of the chemists who perform the syntheses. Spin-labeled lipids seem to have greatly advanced the study of the molecular organization and mobility of membrane lipids with respect to problems of ion transport and membrane potentials. Spin labels based on inhibitors and disrupters of enzymatic processes, including electron transport and oxidative phosphorylation, such as spin-labeled dinitrophenol [555], can provide valuable information on such factors as the structure and geometry of the active centers of enzymes, the nature of the bonds between enzymes and membranes, the topography of enzyme assemblages, and allosteric interactions.

The reduction of nitroxide radicals in biological membranes limits the possibilities of using them to study conformational changes. However, thanks to their electron-acceptor properties, they can be used to study the electron donor–acceptor reactions in the cell.

Hopefully, nitroxide radicals that successfully combine the properties of a structure-sensitive probe and a redox probe will be widely employed in the study of the structure and functions of biological membranes.

NUCLEIC ACIDS

The application of nitroxide radicals to the study of the structure and conformational changes in nucleic acids may be based on the principles presented in Chapters 1–7. Spin labels can be covalently bound to various portions of the macromolecules and provide information on the topography of the matrix in the vicinity of the addition, on the changes in this topography under certain effects, and on the distances between labels [556–570]. The possibilities of noncovalently bound probes were demonstrated in the first paper dealing with the study of nucleic acids with nitroxide radicals [571]. However, at the present time the use of spin labels is still being greatly hampered by a number of specific features of the nucleic acids, primarily by difficulties in the selective chemical modification and by the lack of accurate X-ray diffraction models, which would make it possible, as in the case of many proteins, to precisely identify the location of the labels. These problems are most strongly felt in the work with DNA and to a lesser degree in the work with RNA and tRNA. Knowledge of the sequence of the bases in tRNA and the presence of a clearly defined aminoacyl fragment and sections with minor nucleotides make tRNA a very convenient object for chemical modification with spin labels.

Preparation of spin-labeled preparations of nucleic acids. The chemically active nucleic acid residues presently available for covalent modification by nitroxide radicals can be divided into the following four groups: (a) purine and pyrimidine bases (adenine, guanine, cytosine, uracil, and thymine, (b) hydroxyl groups on the ribose and glucosyl rings, (c) minor bases (e.g., thioruacil), and (d) residues of compounds incorporated into nuclei acids (e.g., aminoacyl residues in tRNA).

The chemical formula of the spin labels that have been proposed for the modification of nucleic acids are given in the accompanying list.

The reactivity of nucleic acids in reactions with spin labels varies considerably as a function of the structure of the object being modified, the nature of the reagent, the medium, and the conditions for carrying out the process. There is a general tendency for separate nucleotides to be modified with greater speed than are nucleotides in macromolecules. The bases in RNA are labeled with much greater ease than are those in DNA, because the reactive groups in the latter are bound by hydrogen bonds and actually hidden within the double helix. The reactivity of individual nucleic acid

Table 16. Values of the Coefficients A and β in the Equation $E = A + \beta \Delta S$ (E and ΔS refer to reactions involving decomposition of enzyme-substrate complexes [293, 338]

Modified Segment	Label	
Bases of polynucleotides, α-NH$_2$ group of aminoacyl in tRNA	(LXIV)	[556, 557]
Bases of polynucleotides, 4-thiouridine tRNA	(XXII)	[558, 559]
Bases of polynucleotides	(LXV)	[556]
Bases of DNA	(LXVI)	[560]
Bases of RNA	(VIII)	[558]
Bases of RNA	(XXIII)	[558]
Polyuridylic acid and tRNA	(LXVII)	[561—563]
Bases of DNA	(LXVIII)	[564, 565]

223

End

Modified Segment	Label		
SH group of cysteine amino-acyl in *t*RNA		(V)	[566]
OH groups of carbohydrate residues of DNA		(LXIX)	[567—569]

derivatives can be estimated from the results of a number of papers in which quantitative measurements were made. Thus, it was found [556] that the reaction of synthetic polynucleotides with spin label LXIV in a 0.1 M phosphate buffer at pH 6.8 and 37° proceeds over the course of 20 hr with the following extents of conversion (in moles of the radical per mole of nucleotide): (a) polycytidylic acid, 10^{-3}, (b) the copolymer of polyuridylic and polyguanylic acids, 3×10^{-4}, and (c) polyadenylic and polyuridylic acids, 10^{-4}. The values of the rate constants for the reaction of various nucleosides with label LXVII at 25° and pH 7 in the presence of 0.1 M N-methylmorpholinium chloride are (in M^{-1} min^{-1}) 260 for uridine, 220 for inosine, and 170 for guanosine [562]. The reactivity of the nucleotides decreases sharply in the transition to the macromolecular state (2 M^{-1} min^{-1} for polyuridylic acid) and even more so after incorporation into *t*RNA (the effective values for the rate constant at the initial moment equal 0.68 M^{-1} min^{-1} for the reaction in water and 0.079 M^{-1} min^{-1} for the reaction in the presence of MgCl$_2$ and NaCl). As *t*RNA is modified by label LXVII, the reactivity of the remaining groups decreases in a regular fashion. As shown in one study [562], the above-mentioned reaction can be used to obtain compounds with a high level of modification by nitroxide radicals.

Separation of the mixture of products following the hydrolysis of yeast RNA labeled with radical XXII at 25° and pH 5.5 in a 0.1 M acetate buffer revealed the following for the extent of conversion of the bases (in %): (a) guanine, 20, (b) adenine, 3, and (c) cytosine and thymine, <1 [558]. At pH 6.5 and 37° alkylating agent LXVI reacts very slowly with calf-thymus DNA in the presence of 0.1 M NaCl. The modification rate increases by a factor of 30–50 in dilute solutions of NaCl (10^{-4} M) when there is some destabilization of the native structure of the polymer [560]. The extent of modification is 10^{-4} to 2×10^{-4} spin labels per nucleotide. The extent of conversion of phage T2 DNA under these conditions is 50–100 times higher than that of the calf-thymus DNA [560].

The possibility of effectively modifying mouse-liver DNA with alkylating agents based on ethylenimine has been demonstrated [564,570]. The DNA preparations were thoroughly purified of RNA hydrolysis products with pancreatic RNAase [564]. The relative reactivity of the bases in the alkylation and acylation reactions varies somewhat as a function of the specific conditions. Although the existing material is still not adequate for serious generalizations, it is clear that guanine and adenine are the most reactive bases and that the reactivity decreases along the following series: separate nucleotides > polynucleotides > RNA > DNA.

The synthesis of the carbonyl imidazole derivative of nitroxide radical LXIX has opened up extensive possibilities for modification [567–569]. According to the data in three studies [567–569], the attack of this reagent is directed at the free hydroxyl groups (2'-OH) of the glucose cytidine residues in DNA. Considering the results due to Kraevskii et al. [572], we can expect that radical LXIX can be used to modify the ribose residues of the respective RNA nucleotides, as well as other ribose derivatives.

The synthesis and study of spin-labeled tRNA preparations is facilitated by the fact that they contain groupings that have an increased reactivity relative to alkylating and acetylating agents, including spin labels XXII and V. A very convenient site for the modification of tRNA is the thiouridine group, which is selectively blocked by label XXII. The specific nature of the binding of label XXII to 4-thiouridine groups was conclusively demonstrated by isolating the corresponding fragment following hydrolysis of spin-labeled tyrosine tRNA [559]. The α-NH$_2$ groups of the aminoacyls in tRNA are preferentially blocked by acylating agent LXIV, making it possible to obtain spin-labeled preparations of tRNA [556]. As shown in one study [566], the aminoacyl residues of tRNA can also be modified at other functional groups. For example, label V selectively modifies the SH group in the cysteine aminoacyl in tRNA. The reaction with this group is completed within 10–15 min, after which there is a considerably slower reaction with other functional RNA residues.

Many workers have noted the possibilities of demodification of spin-labeled preparations due to hydrolysis. For example, tRNA modified with spin label LXVII splits off a radical-containing fragment almost completely at pH 9 within 25–30 min [562]. The ester linkage in label LXVIII hydrolyzes in 40 hr at 40° and neutral pH [564].

The following is a list of the presently known spin-labeled nucleic acid preparations:

Calf thymus	DNA (base)	(LXVI)	[560]
Phage T2	DNA (base)	(LXVI)	[560]
	DNA (glucose–OH)	(LXIX)	[567–569]
Mouse liver,	DNA (base)	(LXVIII)	[564]
normal and	Poly$G$ (base)	(XXII)	[558]
hepatomatous	Poly$A$ (base)	(XXII)	[558]
	Poly$U$ (base)	(XXII)	[558]
	Poly$A$ (base)	(LXIV)	[556]
	Poly$U$ (base)	(LXIV)	[556]
	Poly$C$ (base)	(LXIV)	[556]
	Poly$U$–G (base)	(LXIV)	[556]
	Poly$U$ (base)	(LXVII)	[562]
	DiU (base)	(LXVII)	[562]
Yeast	RNA (base)	(XXII)	[558]
	RNA (base)	(VIII)	[558]
	RNA (base)	(VIII)	[558]
	tRNA (base)	(LXVII)	[562]
	tRNA (base)	(LXVII)	[556]
	tRNA (base)	(V)	[566]
Tyrosine tRNA (4–thioruidine S)		(XXII)	[559]
Methionine tRNA (4–thioruidine S)		(XXII)	[559]
Valine tRNA (4–thioruidine S)		(XXII)	[559]
Phenylalanine tRNA (4–thiouridine S)		(XXII)	[559]
Valine tRNA (aminoacyl α–NH$_2$)		(LXIV)	[557]
Cysteine tRNA (aminoacyl α–NH$_2$)		(V)	[566]

Note. A, adenylic acid; G, guanylic acid; C, cytidylic acid; U, uridylic acid.

Investigation of the microstructure and conformational changes in nucleic acids with spin labels. The high sensitivity of the EPR spectral parameters of nitroxide radicals to the state of the surroundings has made it possible to use spin labels to study the local topography of matrices in the vicinity of bound labels and to monitor conformational changes in nucleic acids as the pH and temperature are varied. As in the experiments with proteins, the addition of spin labels to nucleic acids results in considerable slowing of the rotational diffusion of the nitroxide fragments. The extent of immobilization is determined considerably by the structure of the nucleic acid, the length and flexibility of the label, and the ambient conditions (pH, temperature, and ionic strength). For example, at pH 7.5 and 25° the values of the rotational diffusion parameter ν (in sec^{-1}) for spin label LXVII are as follows [562,563]: (a) 10^{10} for the free label in solution, (b) 6×10^9 on the dimer of uridylic acid, (c) 1.7×10^9 on polyuridylic acid, and (d) 10^9 on tRNA. Different spin labels added to tRNA have different values of ν. For example, this parameter equals 1.5×10^9 sec^{-1} for label LXIV and 10^9 sec^{-1} for label LXVII.

The hydrolysis of yeast RNA by pancreatic ribonuclease [558] is accompanied by an increase in the rotation frequency of label XXII from 2.5×10^9 sec^{-1} to 5×10^9 sec^{-1}. Similarly, at 0° and pH 6.8 the hydrolysis of phenylalanine tRNA results in an increase in ν for label LXIV from 1.3×10^9 sec^{-1} to 3×10^9 sec^{-1} [556]. At the same time the EPR spectra clearly show isobectic points, indicating the absence of intermediate forms with different EPR spectra. The EPR spectra of spin labels, however, are insensitive to the degree of polymerization of the nucleic acids at fairly high molecular weights [558].

Varying the pH in the region for changes in the secondary structure of nucleic acids has a marked effect of the spectral parameters of spin-labeled preparations. At pH 10.95–11.60 polyguanylic acid labeled with radical XXII undergoes a transition that is clearly detected in the EPR spectra of the label [558]. The value of the pK for this transition, which equals 11.5, is in good agreement with the value 11.43 obtained by an optical method. Similarly, the EPR spectra of spin-labeled polyadenylic acid are transformed at pH 5.5–6.5 (the pK of the transition is 5.9).

As already noted, modification of DNA by alkylating spin labels involves certain difficulties apparently due to the blocking of the chemically active residues of the bases by hydrogen bonds. Nevertheless Artyukh et al. [560] have succeeded in carrying out the alkylation of phage T2 and calf-thymus DNA with radical LXVI. The EPR spectra in both cases correspond to an unimmobilized radical with $\nu = 10^{10}$ sec^{-1} regardless of the type of DNA. The binding of approximately one label for every 1000 nucleotides in phage T2 DNA has practically no effect on the value of the hyperchromic effect and the melting point of the preparation. However, modification of 1% of the nucleotides decreases the hyperchromic effect from 38% to 33% and reduces the melting point from 82.5° to 61°. In the case of calf-thymus DNA, disruption of the structure begins even when the extent of modification corresponds to one label for every 5000–10,000 nucleotides. The weak dependence of the value of ν on the structure of the polymer, the high rotation rate of the radical, the increase in the extent of binding under conditions of partial destabilization (low NaCl concentrations), and the loss of the native properties of DNA at high extents of conversion indicate that alkylating agent LXVI is added primarily to bases in despiralized portions, where the motion of the matrix is fairly intense. Nevertheless the mobility of the spin label is very sensitive to pH changes. For example, in the case of sin-labeled phage T2 DNA, the value of ν increases from 2×10^{10} sec^{-1} to 10^{11} sec^{-1} with progression from alkaline to acid pH in the vicinity of pH 2.95. According to the spectrophotometric data, the transition region for the unmodified preparation lies in the 2.55–2.75 pH range.

The use of spin labels based on the alkylating ethylenimine group has opened up interesting possibilities for studying DNA [564,570]. As shown in one study [564], the addition of label LXVIII to mouse-liver DNA results in considerable immobilization of the nitroxide fragment ($\nu \sim 2 \times 10^9$ sec^{-1}). The rotational diffusion parameters of the label on DNA are sensitive to the structure of the macromolecule and vary significantly as a function of the source (Table 17).

Table 17.   Rotational Diffusion Parameters (ν_+, ν_-, and ϵ) of Spin Label LXVII Added to Various Nucleic Acids [564]

Biopolymer	$\nu_+ \cdot 10^9$, sec^{-1}	$\nu^- \cdot 10^9$, sec^{-1}	ϵ
DNA			
mouse liver	3.97	2.16	0.033
hepatoma 22[a]	4.57	2.02	0.027
hepatoma 60	1.08	2.72	0.158
hepatoma 46	1.45	3.94	0.167
Uridylic acid	6.69	12;85	0.118

Let us consider the fact that the parameters ν_+ and ϵ are much more sensitive to the biopolymer structure than is ν_-. Precisely as in the case of the proteins (see Chapters 1 and 7), this situation is due to the anisotropic rotation of the nitroxide fragments and the high sensitivity of the motion of the "legs" of the labels to the structure of the matrix in the vicinity of the addition.

There have been results [567–569] demonstrating the possibility of modifying the hydroxyl groups of the sugar moieties (apparently, the OH groups of glucose) in calf-thymus DNA with label LXIX that are of great importance for the further development of spin-labeling for nucleic acids. This modification has no effect on the system of hydrogen bonds between the bases and probably alters the native structure of the system under investigation to a very small degree. The EPR spectra of label LXIX on DNA indicate that the level of immobilization of the nitroxide fragment is low ($\nu \sim 10^{10}$ sec^{-1}).

Modification of one out of 50 nucleotide pairs has practically no effect on the value of the hyperchromic effect, the transition temperature or width, or the matrix activity of DNA [568]. Complexation of the DNA with RNA polymerase at 37° causes a twofold increase in the rotation frequency of the spin label. However, the EPR spectrum of the complex of DNA with RNA polymerase obtained at 2° does not differ from the spectrum of free DNA. Complexes obtained at different temperatures interact differently with nitrocellulose filters and with certain physiologically active compounds. The high-temperature complex passes freely through the filters and is resistant to the action of heparin, rhythampicin, and high ionic strengths. The complex obtained at 2° is held tightly on the filter and is less stable to the effects just cited. These results indicate that the interaction between DNA and the enzyme RNA polymerase at 37° involves changes in the structure of the polynucleotide. Moreover spin label III bound to the enzyme is significantly immobilized when the complex forms [569] (the value of ν changes from 5×10^8 to 1.6×10^8 sec^{-1}). Addition of a σ subunit and a small amount of the enzyme does not cause noticeable changes in the EPR spectrum of the spin-labeled DNA. All this demonstrates the possibility of detecting fine changes in the structure of nucleic acids when they are functioning.

The spin–spin effects arising in the case of spin-labeled polyuridylic acid (label LXVII) were analyzed in one study [563]. First a spin-labeled fragment consisting of two nucleotides (S–U–U–S) was investigated. At a temperature of 20° the EPR spectrum corresponded to some degree of immobilization ($\nu \sim 6 \times 10^9$ sec^{-1}) of the radical. When the temperature was increased to 85° the spectrum showed additional components indicative of spin exchange during encounters between nitroxide fragments (see Chapter III). The exchange frequency was estimated at $\nu_{\mathrm{exc}} \sim 8 \times 10^8$ sec^{-1}. The exchange interactions manifested themselves clearly in the EPR spectra of the polyuridylic acid. At 20° the spectral lines were broadened significantly, and the size of the hyperfine structure increased as a result of a combination of exchange, anisotropic, and dipole–dipole interactions. Increasing the temperature to 85° enhanced the exchange interactions and weakened the other effects, and the EPR spectrum was transformed into a singlet corresponding to an exchange frequency approximately equal to 7×10^8 sec^{-1}. Evaluation of the local concentration of spins at 77°K made it possible to calculate the value of the exchange rate constant k. It turned out to be equal to 3.3×10^9 M^{-1} sec^{-1} or one third of the value of k for radical LXVII in solution. Thus the study of spin–spin interactions between labels on nucleic acids, as in the case of proteins, yields important additional information concerning the relative positions and type of motion of the radicals, both of which ultimately depend on the structure of the matrix under investigation.

A comparative study of local sections on the boundary between the dihydrouridine and aminoacyl fragments of different types of *t*RNA from *E. coli* was made possible by the addition of spin label XXII to the SH group of 4-thiouridine [559]. This modification has practically no effect on the ability of tyrosine and methionine *t*RNA to accept amino acids. The values of ν calculated from certain data [559] are (in sec^{-1}): (a) $1.3 \times$

10^9 for methionine tRNA, (b) 1.3×10^9 for phenylalanine tRNA, (c) 2×10^9 for tyrosine tRNA, and (d) 4.8×10^9 for valine tRNA. Since the differences between the molecular weights of different types of tRNA are much smaller than the differences between the values of ν, the latter are clearly due to differences in the microstructure. Valine tRNA thus has a more open segment in the vicinity of the 4-thiouridine residue.

The stabilization of the aminoacyl residues modified by label LXIV in tRNA to hydrolysis of the ester bond has made it possible to obtain preparations in which the spin labels are located only on the α-NH$_2$ groups of the incorporated amino acids [556]. This, in turn, has made it possible to monitor local conformational changes in the vicinity of the added labels. Plots of log τ as a function of $1/T$ for spin-labeled valine tRNA show a distinct discontinuity at 51°. In the low-temperature region the effective values of the energy (E_{eff}) and entropy ($\Delta S_{eff}^{\neq}$) for rotational diffusion calculated from the data due to Shofield et al. [556] are 5.3 kcal and 2.5 e.u., respectively. These values fall in the range characteristic of fairly rigid matrices. The values of these parameters for the high-temperature region ($E_{eff} = 10$ kcal and $\Delta S_{eff}^{\neq} = 16$ e.u.) suggest that the matrix has gone into a viscous glycerol-like state. The value of the transition temperature obtained spectrophotometrically is close to the corresponding value obtained from the temperature dependence of ν. This suggests that there is a local change in the vicinity of the aminoacyl active center and the overall restructuring of tRNA coincide. However, no such coincidence was observed upon addition of denaturing agents such as urea and dimethylsulfoxide.

The values of the transition temperatures and E_{eff} varied somewhat as a function of the reaction conditions, the ionic strength, and the RNA structure, but the qualitative laws were identical. Moreover the values of E_{eff} for different samples ranged from 1.6 to 10 kcal, and the values of $\Delta S_{eff}^{\neq}$ ranged from -4.6 to $+37$ e.u., there being an approximately linear correlation (compensation effect) between E_{eff} and $\Delta S_{eff}^{\neq}$. A similar discontinuity has been observed on the log τ versus $1/T$ plots for a preparation of cysteine aminoacyl in tRNA with label V bound to the cysteine SH groups.

Probing the surface of nucleic acids with nitroxide radicals. Freely bound probes are widely used to study the structure and local mobility of condensed media (Chapter 7). They were first used to solve problems in molecular biology in the work of Sukhorukov et al. [571], who studied the state of the surface of nucleic acids under various conditions. The radical probe

$$\text{HO} \quad \text{N—O}^{\cdot} \qquad (10.9)$$

was introduced into dry samples of calf thymus DNA and yeast tRNA in ratios of one probe for every 30 nucleotides. The dry samples showed practically no differences in the shape of their EPR spectra. However, the spectra of the radical were sensitive to the structure of the biopolymers if the measurements were made at different moisture contents.

The first portions of water to be added to the dry samples had no effect of the spectral parameters, which indicated that the radical was strongly immobilized ($\nu < 10^7$ sec^{-1}). After the addition of two water molecules for every nucleotide, the motion of the radicals in the native DNA (nDNA) was greatly restored ($\nu \sim 2 \times 10^9$ sec^{-1}), and further moistening restored even more of its motion. In order to achieve the corresponding restoration of the motion of the probes in denatured DNA (dDNA), it is necessary to add five molecules of water per nucleotide, and in the case of tRNA three molecules are needed. In all of the samples the mobility of the matrix increases approximately linearly with the number of adsorbed water molecules (n); therefore, these workers proposed the quantity $\delta \nu R / \delta n_n$, which they called the specific lability, as a convenient parameter. The value of this parameter for the biopolymers investigated decreases along the series tRNA $> n$DNA $> d$DNA. It was noted that tRNA binds less water than DNA, but the bound water in the former more strongly alters the structure of the polymer in favor of more mobile conformations. The foregoing experiments clearly demonstrate the high sensitivity of the parameters of the EPR spectra of spin probes to the structure of the surface of nucleic acids, as well as the importance of water in the formation of this structure.

Paramagnetic probes have also made it possible to reveal a number of interesting features of the temperature dependence of ν in samples of calf thymus DNA with different water contents. Sukhorukov et al. [573] isolated three approximately linear segments, which they termed segments I ($100°$ to $65°$), II ($65°$ to $20°$), and III ($20°$ to $-30°$), on the Arrhenius plots for the rotational diffusion of radical LXX with a relative moisture content of 80% in the samples.

When the moisture content is higher, the low-temperature region shows a fourth segment, namely, segment IV (below $-13°$ when the water content is 90%, and below $-3°$ when the water content is 100%). The discontinuities on the Arrhenius plots are attributed to structural changes in the DNA–water system. One of these changes was detected in the low-temperature region by the microcalorimetric method [574].

Comparison of the values of the effective activation parameters for rotational diffusion of probe LXX presented in one study [573] with the data in the table on p. 228 makes it possible to discuss the local mobility of the medium in the vicinity of the addition of a probe. In the case of segment IV the values $E_{eff} = 15$ kcal and $\nu_0 = 10^{22}$ sec^{-1} fit the viscous glycerol-like media. Segments II and III seem to be more rigid, and their structure is less yielding to the effect of the temperature (the E_{eff} equal, respectively, 5.1 and 1.8 kcal, and the ν_0 equal 3.7×10^{13} and 1.3×10^{11} sec^{-1}). Finally in segment I the structure of the surface is effectively softened on heating. The structural changes at 66°, 20°, $-4°$, and $-14°$ have been detected in DNA from different sources with the aid of spin labels based on ethylenime by Mil' et al. [570]. These workers noted the existence of a stable conformation in the 20° to 66° temperature range. The values of E_{eff} are sensitive to the nature of the DNA (calf thymus, rat spleen, and tumors).

In conclusion, the first work on the nucleic acids showed that different portions of the nucleic acids can be modified by spin labels and probes despite the well-known difficulties. The development of the chemistry of nucleic acids [575,576] leads us to believe that in the near future the assortment of spin labels and probes will be greatly expanded and that the use of this method in the further study of the structure and conformational changes of the nucleic acids will be stimulated. At the same time, as shown in the pioneer work of McConnell and Ohnishi [454] on the incorporation of a radical ion of chlorpromazine into DNA, other types of paramagnetic species can be used along with nitroxide radicals as spin labels.

Conclusion

During the 8 years of its existence, the spin-label technique has not only demonstrated fundamentally new possibilities for the study of the structure of proteins, enzymes, membranes, nucleic acids, and other biological systems, as well as the conformational changes in them, but has also yielded a whole series of specific important results.

Modification of individual portions of biological structures with nitroxide radicals and paramagnetic ions provides a way to evaluate distances between modified groups in the 2–60 Å range and to determine the depth of paramagnetic centers relative to matrix surfaces. The range is expanded to 100 Å or more by using electron-scattering labels.

Spin labels can serve as specific microscopic seismic stations that detect very small conformational changes in the region where they are bound. A detailed quantitative study of the rotational diffusion of spin labels and probes of different lengths and flexibilities in a biological matrix and the determination of the effectiveness with which they interact with freely diffusing probes provides information on the local topography and mobility of the matrix and on the presence of electrostatic charges near the labels. Analysis of the EPR spectra of nitroxide radicals makes it possible to determine the molecular and conformational mobility of the surrounding medium in the $10^7 - 5 \times 10^{10}$ sec^{-1} frequency range. The use of Mössbauer labels extends the lower limit of this range to 5×10^6 sec^{-1}.

The most important results obtained with the aid of spin labels bound to proteins and enzymes include: (a) the detection and study of the detailed mechanism of allosteric transitions in hemoglobins and aspartate carbamylase, (b) the discovery of cooperative conformational changes in lysozyme, myoglobin, myosin, AAT, and α-chymotrypsin, (c) demonstration of the mobility of different portions of the active center of α-chymotrypsin, (d) proof of the multinuclear nature of the nonheme iron-containing active centers of nitrogenase and its components, ferredoxin, and model proteins, (e) study of the active centers of creatine kinase, dehydrogenase, phosphorylase b, and nitrogenase, and (f) evaluation of the rigidity and local viscosity of the layers in water–protein matrices of enzymes, including the active centers of α-chymotrypsin, myoglobin, and ribonuclease A.

A number of important results concerning the structure of biological membranes have been obtained with the aid of spin probes.

1. The results of the comparative study of the EPR spectra of spin-labeled model lipid membranes, erythrocyte membranes, and nerve-fiber mem-

233

branes carried out by McConnell and his co-workers indicate that there are lipid bilayers in the membranes.

2. Gradients in the mobility of lipids in membranes have been observed. The mobility (flexibility) is very high in the vicinity of the terminal methyl groups of the hydrocarbon chains and considerably lower in the vicinity of the polar tips of the lipids.

3. The high values of the rotational correlaion times and coefficients of lateral diffusion of spin-labeled lipids, steroids, and fatty acids in model and natural membranes underlies the theories on the fluid nature of the hydrocarbon portions of membranes. According to the EPR data, the viscosity of the hydrocarbon portions of biomembranes is about 100 times greater than the viscosity of water.

4. The EPR spectra of spin probes depend on the chemical nature of the lipids in membranes (in particular, on the concentration of unsaturated hydrocarbon chains in the lipids). The spectra depend on the state of the membrane proteins, the temperature, pH, additions of Ca^{2+}, Mg^{2+}, La^{3+}, K^+, and other ions, local anesthetics, drugs, and other membrane effectors.

5. Cooperative structural changes of functional significance have been discovered in electron-transferring membranes of chloroplasts, SMP, and several other membranes with water contents of 10 and 20% in the membranes.

6. Nitroxide derivatives of fatty acids and lipids have been successfully used to detect energy-dependent cooperative conformational changes in mitochondrial membranes and membranes of submitochondrial particles.

The difficulties involved in the chemical modification of nucleic acids with nitroxide labels are being successfully overcome at the present time. Spin labels and probes have been used to monitor conformational changes in DNA and *t*RNA at varying temperatures and pH values and in the presence of denaturing agents. Changes in the structure of the phage T2 DNA macromolecule during the formation of a complex with DNA polymerase have been demonstrated.

There is every reason to assume that the creative potential of the method has not been fully revealed. The chemical systhesis of new spin labels and probes, as well as of mono-, bi-, and polyradicals will open up various possibilities. The incorporation of labels into biological structures by means of biosynthesis appears very promising. New variants of the method can be created by taking advantage of paramagnetic relaxation. Relaxational methods may be able to resolve the question of the "slow" motions of paramagnetic groups with frequencies less than 10^7 sec^{-1} and appreciably expand the range of measurable distances between labels. Vast and un-tapped possibilities may lie in the field of the NMR of spin-labeling systems (analysis of linewidths, relaxation characteristics, contact and pseudo-contact shifts [577–579]), as well as in the combined application of spin, luminescent, Mössbauer, and electron-scattering labels.

Spin-labeling methods, together with other experimental and theoretical techniques, may aid in establishing the tertiary and quaternary structures of enzymes from a known sequence of amino acids without the need for direct X-ray diffraction investigations of single-crystal preparations. In fact, the theoretical techniques presently being developed [580–583] make it possible to predict the secondary structure of a protein from the primary structure. Presumably it will be possible in the near future to theoretically select a certain number of most probable tertiary structures. The final selection among the probable structures may be made by measuring the distances between spin, luminescent, and electron-scattering labels im-planted at known positions along the primary structure. Analysis of the relaxation characteristics and the NMR spectra of paramagnetic nuclei in substrates and enzymes may provide additional information on the dis-tances between spatially distant protein groups.

Obviously the spin-labelling technique has, along with its accomplish-ments, a number of limitations and drawbacks, which are primarily due to the danger of destroying the native form of biological structures upon introduction of the labels. The information obtained with the aid of this method is often of an indirect nature, and the accuracy of the determination of distances by this method falls short of the accuracy of X-ray diffraction analysis. The possibilities of specifically modifying biological systems are also limited. Therefore, the application of the method requires a cautious qualified approach that includes monitoring of the biological activity and the macromolecular structure of the system under study, the use of other chemical and physical methods, and so forth. Nevertheless the material cited above indicates that in some cases spin labels offer unique information on the structure and properties of biological systems and that in other cases spin labeling is a useful addition to the traditional biochemical and bio-physical techniques.

It is thus our hope that spin labeling will continue to be an effective tool for solving various complicated problems in molecular biology.

Bibliography*

1. C. L. Hamilton and H. M. McConnell, in A. Rich and N. Davidson, Editors, *Structural Chemistry and Molecular Biology*, San Francisco (1968), p. 115.
2. H. M. McConnell and B. G. McFarland, *Quart. Rev. Biophys.*, **3**, 91 (1970).
3. G. I. Likhtenshtein, *Usp. Biol. Khim.*, **12**, 3 (1971).
4. V. K. Kol'tover and G. I. Likhtenshtein, *Itogi Nauki, Ser. Molekul. Biol.* (1973).
5. P. Jost and O. H. Griffith, in Colin Chignell, Editor, *Methods in Pharmacology*, Vol. 2, Chap. 7 (1972), p. 223.
6. P. Jost, A. S. Waggoner, and O. H. Griffith, in *Structure and Function of Biological Membranes*, Academic, New York and London (1971), p. 83.
7. I. C. P. Smith, in J. R. Bolton, D. Borg, and H. Schwarz, Editors, *Biological Application of ESR Spectroscopy*, Wiley-Interscience, New York (1971).
8. G. I. Likhtenshtein, in *Methods for Investigating the Photosynthetic Apparatus* [in Russian], Nauka, Moscow (1973).
9. G. I. Likhtenshtein, in *Choline Receptors* [in Russian], Nauka, Moscow (1973).
10. A. N. Kuznetsov and V. A. Livshits, *Chem. Phys. Lett.*, **20**, 535 (1973).
11. C. Weber and F. G. Teal, *Disc. Faraday Soc.*, **27**, 134 (1957).
12. G. I. Likhtenshtein, A. P. Pivovarov, and Yu. N. Smolina, *Molekul. Biol.*, **2**, 291 (1968).
13. E. Ya. Alfimova and G. I. Likhtenshtein, *Biofizika*, **17**, 49 (1972).
14. E. N. Frolov, G. I. Likhtenshtein, and L. A. Syrtsova, *Dokl. Akad. Nauk SSSR*, **196**, 1149 (1971).
15. E. N. Frolov, A. P. Mokrushin, G. I. Likhtenshtein, V. A. Trukhtanov, and V. I. Gol'danskii, *Fourth International Biophysics Congress, Section Reports* [in Russian], Vol. 4, Moscow (1972), p. 297.
16. E. N. Frolov, A. P. Mokrushin, G. I. Likhtenshtein, V. A. Trukhtanov, and V. I. Gol'danskii, *Dokl. Akad. Nauk SSSR*, **212**, 165 (1973).
17. L. A. Levchenko, A. V. Raveskii, A. P. Sadkov, and G. I. Likhtenshtein, *Dokl. Akad. Nauk SSSR*, **211**, 238 (1973).
18. G. I. Likhtenshtein, L. A. Levchenko, A. P. Sadkov, A. V. Raevskii, T. S. Pivovarova, and R. I. Gvozdev, *Dokl. Akad. Nauk SSSR*, **213**, 1442 (1973).
19. G. I. Likhtenshtein, *Molekul. Biol.*, **2**, 234 (1968).
20. G. I. Likhtenshtein and P. Kh. Bobodzhanov, *Biofizika*, **8**, 745 (1968).
21. J. C. Taylor, J. S. Leigh, and M. Cohn, *Proc. Natl. Acad. Sci. U. S.*, **64**, 219 (1969).
22. A. V. Kulikov, G. I. Likhtenshtein, É. G. Rozantsev, V. I. Suskina, and A. B. Shapiro, *Biofizika*, **17**, 42 (1972).
23. A. I. Kokorin, K. I. Zamarev, G. L. Grigoryan, V. P. Ivanov, and É. G. Rozantsev, *Biofizika*, **17**, 34 (1972).
24. A. V. Kulikov and G. I. Likhtenshtein, *Biofizika*, **19**, 420 (1974).
25. G. I. Likhtenshtein, Yu. B. Grebenshchikov, and T. V. Avilova, *Molekul. Biol.*, **6**, 67 (1972).
26. D. E. Kosman, J. C. Hsia, and L. H. Piette, *Arch. Biochem. Biophys.*, **133**, 29 (1969).

*For the names of the Russian periodicals available in English translation, see the list at the end of the bibliography.

27. E. N. Frolov, G. I. Likhtenshtein, Yu. I. Khurgin, and O. V. Belonogova, *Izv. Akad. Nauk SSSR, Ser. Khim.*, No. 1, 132 (1973).
28. S. Ogawa, and H. M. McConnell, *Proc. Natl. Acad. Sci. U S.*, **58**, 19 (1967).
29. R. T. Ogata and H. M. McConnell, *Proc. Natl. Acad. Sci. U. S.*, **69**, 335 (1972).
30. G. I. Likhtenshtein and Yu. D. Akhmedov, *Molekul. Biol.*, **4**, 551 (1970).
31. Yu. B. Grebenshchikov, G. G. Charkviani, N. I. Gachechiladze, Yu. V. Kokhanov, and G. I. Likhtenshtein, *Biofizika*, **17**, 794 (1972).
32. B. P. Atanasov, Ju. D. Achmedov, and G. I. Likhtenshtein (in press).
33. V. P. Timofeev, O. L. Polyanovskii, M. V. Vol'kenshtein, G. I. Likhtenshtein, and Yu. V. Kokhanov, *Molekul. Biol.*, **6**, 373 (1972).
34. V. P. Timofeev, O. L. Polyanovskii, M. V. Vol'kenshtein, and G. I. Likhtenshtein, *Biochim. Biophys. Acta*, **220**, 357 (1970).
35. M. B. Neiman, E. G. Rosantsev, and Ju. G. Mamedova, *Nature*, **196**, 472 (1962).
36. É. G. Rozantsev, *Free Nitroxide Radicals* [in Russian], Khimiya, Moscow (1970).
37. A. L. Buchachenko, *Stable Radicals* [in Russian], Izd-vo AN SSSR, Moscow (1963).
38. A. L. Buchachenko and A. M. Vasserman, *Stable Radicals* [in Russian], Khimiya, Moscow (1973).
39. D. Kivelson, *J. Chem. Phys.*, **33**, 1094 (1960).
40. H. M. McConnell, *J. Chem. Phys.*, **25**, 709 (1956).
41. T. J. Stone, T. Buchman, P. L. Nordio, and H. M. McConnell, *Proc. Natl. Acad. Sci. U. S.*, **54**, 1010 (1965).
42. G. I. Likhtenshtein, Yu. B. Grebenshchikov, P. Kh. Bobodzhanov, and Yu. V. Kokhanov, *Molekul. Biol.*, **4**, 482 (1970).
43. G. I. Likhtenshtein, *Fourth International Biophysics Congress, Symposium Reports* [in Russian], Nauka, Moscow (1973).
44. E. N. Frolov, G. I. Likhtenshtein, and N. V. Kharakhonycheva, *Molekul. Biol.*, **8**, 886 (1973).
45. G. I. Likhtenshtein, Yu. B. Grebenshchikov, T. V. Troshkina, E. N. Frolov, Yu. D. Akhmedov, and A. V. Kulikov, *Seventh International Symposium on the Chemistry of Natural Compounds, Reports* [in Russian], Zinatne, Riga (1970), p. 714.
46. L. A. Syrtsova, L. A. Levchenko, E. N. Frolov, G. I. Likhtenshtein, T. N. Pisarskaya, L. V. Vorob'ev, and V. A. Gromoglasova, *Molekul. Biol.*, **5**, 726 (1971).
47. A. V. Kulikov, L. A. Syrtsova, G. I. Likhtenshtein, and T. N. Pisarskaya, *Molekul. Biol.*, **9**, 203 (1975).
48. V. K. Kol'tover, Yu. A. Kutlakhmedov, and B. I. Sukhorukov, *Dokl. Akad. Nauk SSSR*, **181**, 730 (1968).
49. E. N. Mukhin, V. K. Gins, E. N. Frolov, and G. I. Likhtenshtein, *Dokl. Akad. Nauk SSSR*, **207**, 235 (1973).
50. A. E. Shilov and G. I. Likhtenshtein, *Izv. Akad. Nauk SSSR, Ser. Biol.*, No. 4, 518 (1971).
51. V. K. Koltover, M. G. Goldfeld, L. Ya. Gendel, and É. G. Rozantsev, *Biochem. Biophys. Res. Commun.*, **32**, 421 (1968).
52. L. I. Antsiferova, A. V. Lazarev, and V. B. Stryukov, *Zh. Éksperim. i Teor. Fiz., Pis'ma*, **12**, 108 (1970).
53. A. V. Lazarev and V. B. Stryukov, *Dokl. Akad. Nauk SSSR*, **197**, 627 (1971).
54. W. L. Hubbell and H. M. McConnell, *J. Amer. Chem. Soc.*, **93**, 314 (1971).
55. S. A. Goldman, G. V. Bruno, C. F. Polnaszek, and J. H. Freed, *J. Chem. Phys.*, **56**, 716 (1972).
56. L. I. Antsiferova, *Zh. Fiz. Khim.*, **47**, 794 (1973).
57. J. H. Freed and G. H. Frenkel, *J. Chem. Phys.*, **39**, 326 (1963).

58. J. H. Freed, G. V. Bruno, and C. F. Polnaszek, *J. Phys. Chem.*, **75**, 3385 (1971).
59. I. V. Alexandrov, A. N. Ivanova, N. N. Korst, A. V. Lazarev, A. I. Prokhozhenko, and V. B. Stryukov, *Molec. Phys.*, **18**, 681 (1970).
60. A. M. Vasserman, A. N. Kuznetsov, A. L. Kovarskii, and A. L. Buchachenko, *Zh. Strukt. Khim.*, **12**, 609 (1971).
61. G. E. Pake, *Paramagnetic Resonance*, Benjamin, New York (1962).
62. N. Edelstein, A. Kwok, and A. H. Maki, *J. Chem. Phys.*, **41**, 179 (1964).
63. A. N. Kuznetsov, A. M. Wasserman, A. Ju. Volkov, and N. N. Korst, *Chem. Phys. Lett.*, **12**, 103 (1971).
64. R. C. McCalley, E. J. Shimshick, and H. M. McConnell, *Chem. Phys. Lett.*, **13**, 115 (1972).
65. E. J. Shimshick and H. M. McConnell, *Biochem. Biophys. Res. Commun.*, **46**, 321 (1972).
66. M. S. Itzkowitz, Ph. D. Thesis, California Institute of Technology (1966).
67. N. N. Korst and A. O. Lazarev, *Molec. Phys.*, **17**, 481 (1969).
68. J. S. Hyde, *Fourth International Biophysics Congress, Symposium Reports* [in Russian], Nauka, Moscow, p. 65.
69. O. H. Griffith and A. S. Waggoner, *Acc. Chem. Res.*, **2**, 17 (1969).
70. N. N. Korst and T. N. Khazanovich, *Zh. Éksperim. i Teor. Fiz.*, **45**, 1523 (1963).
71. I. V. Aleksandrov, *Magnetic Resonance* [in Russian], Nauka, Moscow (1964).
72. S. A. Goldman, G. V. Bruno, and J. H. Freed, *J. Phys. Chem.*, **76**, 1858 (1972).
73. N. N. Korst, *Teor. i Matem. Fiz.*, **6**, 265 (1971).
74. H. Silescu, *J. Chem. Phys.*, **54**, 2110 (1971).
75. R. Kubo and K. Tomita, *J. Chem. Phys. Soc. Japan*, **9**, 888 (1956).
76. Yu. S. Karimov, *Zh. Éksperim. i Teor. Fiz. Pis'ma*, **8**, 239 (1968).
77. S. T. Kirillov, *Motion and Paramagnetic Relaxation of Nitroxide Radicals in Viscous Media* [in Russian], Diploma Piece, Institute of Chemical Physics, USSR Academy of Sciences, Moscow (1973).
78. M. Itzkowitz, *J. Chem. Phys.*, **46**, 3047 (1967).
79. V. B. Stryukov, *Application of Nitroxide Radicals to the Study of Polymers* [in Russian], Master's Dissertation, Moscow (1968).
80. D. Wallach, *J. Chem. Phys.*, **47**, 5258 (1967).
81. G. I. Likhtenshtein, *Role of Conformational and Energy Factors in Enzymatic Catalysis* [in Russian], Doctoral Dissertation, Moscow (1972).
82. T. V. Avilova, *Kinetic Features of Conformational Changes in Proteins* [in Russian], Master's Dissertation, Moscow (1972).
83. Yu. D. Akhmedov, *Investigation of Conformational Changes in Proteins and Enzymes with the Aid of Spin Labels* [in Russian], Master's Dissertation, Moscow (1972).
84. G. I. Likhtenshtein, T. V. Troshkina, Yu. D. Akhmedov, and V. F. Shuvalov, *Molekul. Biol.*, **3**, 413 (1969).
85. G. I. Likhtenshtein, *Studia Biophys.*, **33**, 185 (1972).
86. G. I. Likhtenshtein, Ju. B. Grebenshikov, T. W. Avilova, E. N. Frolov, and J. D. Achmedov, *Abstracts of the Seventh Meeting of FEBS*, Varna (1971), p. 109.
87. G. I. Likhtenshtein, T. V. Avilova, Yu. B. Grebenshchikov, and L. V. Ivanov, *Fourth International Biophysics Congress, Symposium Reports* [in Russian], Vol. 2, Moscow (1972), p. 54.
88. G. I. Likhtenshtein, T. V. Avilova, E. N. Frolov, L. V. Ivanov, O. V. Belonogova, and Yu. I. Khurgin, *All-Union Seminar on Conformational Changes in Biopolymers in Solution, Theses* [in Russian], Metsniereba, Tbilisi (1973).
89. O. L. Lebedev, M. L. Khidekel', and G. A. Razuvaev, *Dokl. Akad. Nauk SSSR*, **140**, 1327 (1961).

90. A. R. Forrester, J. M. Hay, and R. H. Thomson, *Organic Chemistry of Stable Free Radicals*, Academic, New York (1968), p. 180.

91. V. A. Golubev, *Synthesis of Stable Nitroxide Radicals* [in Russian], Master's Dissertation, Moscow (1964).

92. P. Kh. Bobodzhanov, *Synthesis and Study of Spin-Labeled Proteins and Enzymes* [in Russian], Master's Dissertation, Dushanbe (1972).

93. G. I. Likhtenshtein, E. N. Frolov, N. F. Neznaiko, L. A. Levchenko, and Yu. S. Sklyar, *Molekul. Biol.*, **6**, 201 (1972).

94. E. N. Frolov, *Investigation of Nonheme Iron- and Sulfur-Containing Proteins* [in Russian], Master's Dissertation, Moscow (1972).

95. O. Griffith and H. M. McConnell, *Proc. Natl. Acad. Sci. U. S.*, **55**, 8 (1966).

95a. J. C. A. Boeyens, H. M. McConnell, *Proc. Natl. Acad. Sci. U.S.*, **56**, 22 (1966).

96. D. S. Markovich, G. B. Postnikova, N. V. Umrikhina, A. I. Artyukh, B. S. Marinov, M. V. Vol'kenshtein, and B. I. Sukhorukov, *Biofizika*, **16**, 1131 (1971).

97. H. M. McConnell, W. Deal, and R. T. Ogata, *Biochemistry*, **8**, 2580 (1969).

97a. H. M. McConnell and C. L. Hamilton, *Biochemistry*, **60**, 776 (1968).

98. Yu. V. Kokhanov, É. G. Rozantsev, L. N. Nikolaenko, and L. A. Maksimova, *Khim. Geterotsikl. Soedin.*, No. 11, 1527 (1971).

99. V. T. Ivanov, L. V. Sumskaya, I. I. Mikhaleva, I. A. Lainen, I. D. Ryabova, and Yu. A. Ovchinnikov, *Khim. Prirodn. Soedin.* (1973) (in press).

100. G. I. Likhtenshtein, A. P. Pivovarov, P. Kh. Bobodzhanov, É. G. Rozantsev, and I. B. Smolina, *Biofizika*, **13**, 396 (1969).

101. G. I. Likhtenshtein, P. Kh. Bobodzhanov, É. G. Rozantsev, and V. I. Suskina, *Molekul. Biol.*, **2**, 44 (1968).

102. R. W. Burley, J. C. Seidel, and J. Gergely, *Arch. Biochem. Biophys.*, **146**, 597 (1971).

103. W. Baltasar, *Eur. J. Biochem.*, **22**, 158 (1971).

104. L. I. Berliner and H. M. McConnell, *Proc. Natl. Acad. Sci. U. S.*, **55**, 708 (1966).

105. J. S. Hsia, D. I. Kosman, and L. H. Piette, *Biochem. Biophys. Res. Commun.*, **36**, 75 (1969).

106. J. D. Morrisett, C. A. Broomfield, and B. E. Hackley, Jr., *J. Biol. Chem.*, **244**, 5758 (1969).

107. V. A. Yakovlev, O. A. Khodakovskaya, and T. P. Gusovskaya, *Biofizika*, **18** (1973).

108. G. L. Grigoryan, *Dokl. Akad. Nauk SSSR*, **201**, 224 (1971).

109. D. J. Kosman, J. C. Hsia, and L. H. Piette, *Arch. Biochem. Biophys.*, **133**, 29 (1969).

110. J. C. P. Smith, *Biochemistry*, **7**, 745 (1968).

111. D. J. Kosman, *J. Molec. Biol.*, **67**, 247 (1972).

112. M. Barrat, G. Dodd, and D. Chapman, *Biochim. Biophys. Acta*, **194**, 600 (1969).

113. G. I. Likhtenshtein, Yu. D. Akhmedov, L. V. Ivanov, and L. A. Krinitskaya, *Molekul. Biol.*, **8**, 48 (1974).

114. Yu. D. Akhmedov, G. I. Likhtenshtein, L. V. Ivanov, and Yu. V. Kokhanov, *Dokl. Akad. Nauk SSSR*, **205**, 372 (1972).

115. P. V. Sergeev, Yu. B. Grebenshchikov, T. I. Ul'yankina, R. D. Seifula, and G. I. Likhtenshtein, *Molekul. Biol.*, **8**, 206 (1974).

116. G. L. Grigoryan, M. M. Korkhmazyan, and É. G. Rozantsev, *Biofizika*, **17**, 692 (1972).

117. H. Weiner, *Biochemistry*, **8**, 526 (1969).

118. A. S. Mildvan and H. Weiner, *Biochemistry*, **8**, 552 (1969).

119. A. S. Mildvan and H. Weiner, *J. Biol. Chem.*, **244**, 2465 (1969).

120. L. H. Piette and J. E. Spollholz, *Fourth International Biophysics Congress, Section Reports* [in Russian], Vol. 1, Moscow (1972), p. 88.

121. R. Cooke and J. Duke, *J. Biol. Chem.*, **246**, 6360 (1971).

122. T. Tokiva, *Biochem. Biophys. Res. Commun.*, **44**, 471 (1971).
123. L. S. Berliner, *J. Molec. Biol.*, **61**, 189 (1971).
124. R. W. Wien, J. D. Morriset, and H. M. McConnell, *Biochemistry*, **11**, 3707 (1972).
125. J. Taylor, P. Mushak, and J. E. Coleman, *Proc. Natl. Acad. Sci. U. S.*, **67**, 1410 (1970).
126. C. F. Chignell and R. H. Erlich, *Fourth International Biophysics Congress, Section Reports* [in Russian], Vol. 1, Moscow (1972), p. 89.
127. A. B. Shapiro, B. V. Rozynov, É. G. Rozantsev, N. F. Kucherova, L. A. Aksanova, and N. N. Novikoya, *Izv. Akad. Nauk SSSR, Ser. Khim.*, No. 4, 867 (1971).
128. C. K. Pobert, S. J. Hannach, and O. Jardetzky, *Science*, **165**, 504 (1969).
129. R. T. Ogata and H. M. McConnell, *Cold Spring Harbor Sympos. Quant. Biol.*, **36**, 325 (1972).
130. L. M. Raikhman and B. P. Annaev, *Molekul. Biol.*, **6**, 557 (1972).
131. V. S. Belova, *Investigation of Hydroxylation Reactions in Model and Biological Systems* [in Russian], Master's Dissertation, Moscow (1972).
132. L. M. Reichman, B. P. Annaev, W. S. Belova, and É. G. Rozanzev, *Nature, New Biol.*, **237**, 31 (1972).
133. B. P. Annaev, V. P. Ivanov, L. M. Raikhman, and É. G. Rozantsev, *Izv. Akad. Nauk SSSR, Ser. Khim.*, No. 12, 2814 (1971).
134. R. T. Ogata, and H. M. McConnell, *Biochemistry*, **11**, 4792 (1972).
135. L. A. Syrtsova, T. N. Pisarskaya, V. L. Berdinskii, G. I. Likhtenshtein, V. P. Lezina, and A. U. Stepanyants, *Molekul. Biol.*, **8**, 824 (1974).
136. J. C. Hsia and L. H. Piette, *Arch. Biochem. Biophys.*, **129**, 296 (1969).
137. L. H. Berliner, *Fourth International Biophysics Congress, Section Reports* [in Russian], Vol. 1, Moscow (1972), p. 86.
138. Yu. M. Torchinskii, *Sulfhydryl Groups and Disulfide Bridges in Proteins* [in Russian], Nauka, Moscow (1971).
139. J. K. Moffat, *J. Molec. Biol.*, **55**, 135 (1971).
140. M. A. Chekalin and A. D. Virnik, *Zh. Prikl. Khim.*, **35**, 588 (1962).
141. I. Ya. Kalantarov, *Active Dyes* [in Russian], Irfon, Dushanbe (1970).
142. E. N. Frolov, G. I. Likhtenshtein, Yu. I. Khurgin, and O. V. Belonogova, *All-Union Conference on the Chemistry of Proteolytic Enzymes, Theses* [in Russian], Izd-vo Pyargale, Vil'nyus (1973), p. 87.
143. N. V. Kharakhonycheva, Ya. A. Sigidin, and G. I. Likhtenshtein, *Farmakologiya i Toksikologiya* (1973).
144. E. N. Mukhin, V. K. Gins, E. N. Frolov, and G. I. Likhtenshtein, *Fourth International Biophysics Congress, Section Reports* [in Russian], Vol. 1, Moscow (1972), p. 322.
145. E. N. Mukhin, V. K. Gins, A. V. Kulikov, and G. I. Likhtenshtein, *Fiziol. Rastenii* (1973).
146. A. I. Kyaivyaryainen, R. S. Nezlin, M. V. Vol'kenshtein, and G. I. Likhtenshtein, *Molekul. Biol.*, **7**, 760 (1973).
147. J. Quinlivan, H. M. McConnell, L. Stowring, R. Coove, and M. F. Morales, *Biochemistry*, **8**, 3644 (1969).
148. J. C. Seidel and J. Gergely, *Biochem. Biophys. Res. Communs.*, **44**, 826 (1971).
149. D. S. Stone, *Arch. Biochem. Biophys.*, **141**, 378 (1970).
150. L. G. Ignat'eva, G. M. Seregina, L. A. Blyumenfel'd, É. K. Ruuge, R. I. Artyukh, and G. B. Postnikova, *Biofizika*, **17**, 533 (1972).
151. R. W. Burley, J. C. Seidel, and J. Gergely, *Arch. Biochem. Biophys.*, **150**, 792 (1972).
152. L. G. Ignat'eva and É. K. Ruuge, *Biofizika*, **17**, 148 (1972).

153. R. W. Burley, J. C. Seidel, and J. Gergely, *Arch. Biochem. Biophys.*, **150,** 792 (1972).
154. H. R. Drott, C. P. Lee, and T. Yonetani, *J. Biol. Chem.*, **245,** 5875 (1970).
155. Yu. B. Grebenshchikov, *Investigation of the Microstructure of Proteins and Enzymes by the Spin Label–Spin Probe Method* [in Russian], Master's Dissertation, Moscow (1973).
156. A. Bennick, J. D. Campbell, R. A. Dwek, N. C. Price, G. K. Rudda, and A. D. Salmon, *Nature, New Biol.*, **234,** 140 (1971).
157. M. Cohn and J. Reuben, *Acc. Chem. Res.*, **4,** 214 (1971).
158. J. S. Leigh, Jr., *J. Chem. Phys.*, **52,** 2608 (1970).
159. Ya. S. Lebedev, *Dokl. Akad. Nauk SSSR*, **171,** 38 (1966).
160. A. Abragam, *The Principles of Nuclear Magnetism*, Oxford, U. P. London (1961).
161. A. V. Kulikov, *Application of Spin Labels to the Determination of Distances between Protein Groups* [in Russian], Diploma Piece, Institute of Chemical Physics, USSR Academy of Sciences (1970).
162. G. M. Zhidomirov and A. L. Buchachenko, *Zh. Strukt. Khim.*, **8,** 1110 (1967).
163. A. L. Buchachenko, in *Free-Radical States in Chemistry* [in Russian], Nauka, Novosibirsk (1972).
164. A. L. Buchachenko, L. V. Ruban, and É. G. Rozantsev, in *Radiospectroscopic and Quantum-Mechanical Methods in Structural Investigations* [in Russian], Nauka, Moscow (1967).
165. É. G. Rozantsev, V. A. Golubev, M. B. Neiman, and Yu. V. Kokhanov, *Izv. Akad. Nauk SSSR, Ser. Khim.*, No. 3, 572 (1965).
166. A. I. Kokorin, *Determination of Distances between Nitroxide Radicals from the Parameters of EPR Spectra* [in Russian], Diploma Piece, Institute of Chemical Physics, USSR Academy of Sciences, Moscow (1970).
167. G. I. Likhtenshtein and A. V. Kulikov, *Fourth International Biophysics Congress, Section Reports* [in Russian], Vol. 1, Moscow (1972), p. 85.
168. A. I. Kotel'nikov, R. I. Gvozdev, A. V. Kulikov, and G. I. Likhtenshtein, *Studia Biophys.*, **49,** 215 (1975).
169. A. V. Kulikov, V. I. Muromtsev, and S. N. Safronov, *Fiz. Tverd. Tela* (1974).
170. A. V. Kulikov and G. I. Likhtenshtein, *Molekul. Biol.*, **10,** N1 (1974).
171. Ya. S. Lebedev and V. I. Muromtsev, *EPR* and *Relaxation of Trapped Radicals* [in Russian], Khimiya, Moscow (1972).
172. O. Ya. Grinberg, A. A. Dubinskii, and Ya. S. Lebedev, *Dokl. Akad. Nauk SSSR*, **196,** 627 (1971).
173. A. I. Nepomnyashchii, V. I. Muromtsev, and Kh. S. Bagdasar'yan, *Dokl. Akad. Nauk SSSR*, **149,** 901 (1963).
174. A. I. Kyaivyaryainen, M. V. Vol'kenshtein, and V. P. Timofeev, *Molekul. Biol.*, **6,** 557 (1972).
175. J. S. Hyde, C. W. Chien, and J. H. Freed, *J. Chem. Phys.*, **48,** 4211 (1968).
176. V. A. Benderskii, L. A. Blumenfeld, P. A. Stunzhas, and E. A. Sokolov, *Nature*, **220,** 365 (1970).
177. V. A. Benderskii, P. A. Stunzhas, and E. A. Sokolov, *Opt. i Spektrosk.*, **28,** 432 (1970).
178. G. I. Giotta and H. W. Wang, *Biochem. Biophys. Res. Communs.*, **46,** 1576 (1972).
179. M. N. Nechtschein and J. S. Hyde, *Phys. Rev. Lett.*, **24,** 672 (1972).
180. V. A. Benderskii, V. I. Gol'danskii, P. A. Stunzhas, E. A. Sokolov, and A. I. Rakoed, *Dokl. Akad. Nauk SSSR*, **204,** 1143 (1972).
181. A Dershamscy, A. Geogrieva, K. Kotev, and B. P. Atanasov, *Biochim. Biophys. Acta*, **214,** 83 (1970).

182. D. Phillips, in *Molecules and Cells* [Russian translation], Mir, Moscow (1968).
183. D. C. Phillips, *Sci. Amer.*, **215**, 78 (1966).
184. B. I. Stepanov and T. I. Vaganova, *Izv. Akad. Nauk SSSR, Ser. Khim.*, No. 12, 2785 (1967).
185. L. V. Kozlov, L. M. Ginodman, and V. N. Orekhovich, *Biokhimiya*, **32**, 1011 (1967).
186. J. C. Kendrew, *Progr. Biophys.*, **4**, 244 (1954).
187. A. G. San Pietro, Editor, *Non-heme Iron Proteins*, Antioch Press, Yellow Springs, Ohio (1965).
188. L. A. Syrtsova and G. I. Likhtenshtein, *Usp. Biol. Khim.*, **11**, 149 (1971).
189. H. Neurath, Editor, "Metalloenzymes," in *The Proteins*, Vol. 5 (1970).
190. A. F. Vanin, L. A. Blumenfel'd and A. G. Chetverikov, *Biofizika*, **12**, 829 (1967).
191. R. W. F. Hardy and R. C. Burns, *Ann. Rev. Biochem.*, **37**, 331 (1968).
192. J. A. Ovchinnikov, V. I. Ivanov, V. F. Bystrov, A. I. Miroshnikov, E. N. Shepel, N. D. Abdullaev, E. S. Efremov, and L. B. Senyavina, *Biochem. Biophys. Res. Commun.*, **39**, 217 (1970).
193. P. De Santis and A. M. Liquori, *Biopolymers*, **10**, 699 (1971).
194. V. T. Ivanov, A. I. Miroshnikov, L. G. Snezhkova, Yu. A. Ovchinnikov, A. V. Kulikov, and G. I. Likhtenshtein, *Khim Prirodn. Soed.*, No. 1, 91 (1973).
195. D. C. Hodgkin and B. M. Oughton, *Biochem., J.*, **65**, 752 (1957).
196. G. E. Pake and T. R. Tuttle, *Phys. Rev. Lett.*, **3**, 423 (1959).
197. G. I. Skubnevskaya and Yu. N. Molin, *Kinetika i Kataliz*, **8**, 1192 (1967).
198. J. Owen, *J. Appl. Phys., Suppl.*, **33**, 355 (1962).
199. K. I. Zamaraev, A. T. Nikitaev, G. A. Senyukova, and A. Ya. Sychev, *All-Union Conference on Magnetic Resonance, Reports* [in Russian], Kazan' (1969).
200. A. L. Buchachenko, A. M. Wasserman, and A. L. Kovarski, *Internat. J. Chem. Kinetics*, **1**, 361 (1969).
201. J. D. Currin, *Phys. Rev.*, **126**, 1995 (1962).
202. G. I. Skubnevskaya, E. E. Zave, R. I. Zusman, and Yu. N. Molin, *Dokl. Akad. Nauk SSSR*, **170**, 386 (1966).
203. M. P. Eastman, R. G. Kooser, M. R. Das, and J. H. Freed, *J. Chem. Phys.*, **51**, 2690 (1969).
204. G. I. Skubnevskaya, K. M. Salikhov, L. N. Smirnova, and Yu. N. Molin, *Kinetika i Kataliz*, **11**, 889 (1970).
205. G. I. Skubnevskaya and Yu. N. Molin, *Kinetika i Kataliz*, **13**, 1383 (1972).
206. K. M. Salikhov, A. T. Nikitaev, G. A. Senyukova, and K. I. Zamaraev, *Teor. i Éksperim. Khim.*, **7**, 619 (1971).
207. O. A. Anisimov, A. T. Nikitaev, and K. I. Zamaraev, *Teor. i Éksperim. Khim.*, **7**, 682 (1971).
208. K. I. Zamaraev, A. T. Nikitaev, and G. A. Senyukova, *Kinetika i Kataliz*, **13**, 54 (1972).
209. K. I. Zamaraev and A. T. Nikitaev, in *International Collection in Memory of V. V. Voevodskii* [in Russian], Nauka, Novosibirsk (1972), p. 102.
210. K. M. Salichov, A. B. Doctorov, Yu. N. Molin, and K. I. Zamaraev, *J. Magnet. Resonance*, **5**, 189 (1971).
211. Yu. B. Grebenshchikov, G. I. Likhtenshtein, V. P. Ivanov, and É. G. Rozantsev, *Molekul. Biol.*, **6**, 498 (1972).
212. A. L. Buchachenko, *Research in the Field of Stable Radicals* [in Russian], Doctoral Dissertation, Moscow (1968).
213. P. Debye, *Trans. Electrochem. Soc.*, **82**, 265 (1942).
214. J. L. Webb, *Enzyme and Metabolic Inhibitors*, Academic, New York (1966).

215. Yu. B. Grebenshchikov, G. V. Ponomarev, R. P. Evstigneeva, and G. I. Likhtenshtein, *Biofizika*, **17**, 910 (1972).

216. M. F. Perutz, *J. Molec. Biol.*, **13**, 646 (1965).

217. H. M. McConnell, W. Deal, and R. T. Ogata, *Biochemistry*, **8**, (6), 2580 (1969).

218. O. Jardetzky, *Adv. Chem. Phys.*, **7**, 499 (1964).

219. O. Jardetzky and N. G. Wade-Jardetzky, *Molec. Pharmacol.*, **1**, 214 (1965).

220. M. A. Raftery, F. W. Dahlquist, S. M. Parsons, and R. G. Wolcott, *Proc. Natl. Acad. Sci. U. S.*, **62**, 44 (1969).

221. J. Reuben, *Proc. Natl. Acad. Sci. U. S.*, **68**, 563 (1971).

222. B. D. Sykes, *Biochemistry*, **8**, 1110 (1969).

223. D. P. Hollis, *Biochemistry*, **6**, 2080 (1967).

224. B. D. Sykes, P. G. Schmidt, and G. R. Stark, *J. Biol. Chem.*, **245**, 1180 (1970).

225. T. Nowak and A. S. Mildvan, *Biochemistry*, **11**, 2813 (1972).

226. V. F. Bystrov, *Zh. Vsesoyuzn. Khim. Ob-va*, **16**, 380 (1971).

227. O. Jardezky, *Fourth International Biophysics Congress, Symposium Reports* [in Russian], Moscow (1972), p. 130.

228. I. Solomon, *Phys. Rev.*, **99**, 559 (1955).

229. I. Solomon and N. Blombergen, *J. Chem. Phys.*, **25**, 261 (1956).

230. N. Blombergen and L. Morgan, *J. Chem. Phys.*, **34**, 842 (1961).

231. A. Lanir and G. Navon, *Biochemistry*, **10**, 1024 (1971).

232. T. Nowak and A. S. Mildvan, *Biochemistry*, **11**, 2819 (1972).

233. L. A. Syrtsova, I. I. Nazarova, V. B. Nazarov, and T. N. Pisarskaya, *Dokl. Akad. Nauk SSSR*, **206**, 367 (1972).

234. B. E. Smith, D. L. Lowe, and R. C. Bray, *Biochem. J.*, **130**, 641 (1972).

235. R. R. Eady, B. E. Smith, K. A. Cook, and J. R. Postgate, *Biochem. J.*, **128**, 655 (1972).

236. G. L. Grigorian, W. I. Suskina, É. G. Rozantzev, and A. E. Kalmanson, *Nature*, **216**, 927 (1967).

237. G. L. Grigoryan, V. I. Suskina, É. G. Rozantsev, and A. E. Kalmanson, *Molekul. Biol.*, **2**, 148 (1968).

238. S. Ohnishi, J. C. Boyens, and H. M. McConnell, *Proc. Natl. Acad. Sci. U S.*, **56**, 809 (1966).

239. S. Ohnishi, T. Moeda, T. Ito, and K. Huang, *Biochemistry*, **7**, 266 (1968).

240. S. Ogawa, H. M. McConnell, and S. Horwitz, *Proc. Natl. Acad. Sci. U. S.*, **61**, 401 (1968).

241. C. Ho, J. J. Baldassare, and S. Charache, *Proc. Natl. Acad. Sci. U. S.*, **66**, 722 (1970).

242. J. J. Baldassare, S. Charache, R. T. Jones, and C. Ho, *Biochemistry*, **9**, 111 (1970).

243. T. Asakura and H. R. Drott, *Biochem. Biophys. Res. Commun.*, **44**, 1199 (1971).

244. T. Buchman, *Biochemistry*, **9**, 3254 (1970).

245. I. A. Rupley, *Proc. Roy. Soc.*, **B167**, 316 (1967).

246. F. W. Danlquist and M. A. Raltery, *Biochemistry*, **7**, 3269 (1969).

247. J. J. Pollock, U. Zehavi, V. Teichlerg, and N. Sharon, *Israel J. Chem.*, **6**, 112 (1968).

248. B. Yu. Zaslavskii, A. A. Vichutinskii, and A. Ya. Khorlin, *Biofizika*, **17**, 412 (1972).

249. B. Yu. Zaslavskii, *Microcalorimetric Investigation of Lysozyme Reactions* [in Russian], Master's Dissertation, Moscow (1972).

250. Yu. B. Grebenshchikov, G. G. Charkviani, N. A. Gachechiladze, Yu. V. Kokhanov, and G. I. Likhtenshtein, *Third All-Union Symposium on the Biophysics of Muscular Contraction*, Theses [in Russian], Erevan (1971), p. 32.

251. G. G. Charkviani, N. A. Gachechiladze, G. I. Likhtenshtein, Yu. B. Grebenshchikov, and A. V. Kulikov, *Republic Scientific Conference of the Junior Scientists of the Georgian SSR*, Theses [in Russian], Tbilisi (1970), p. 131.

252. M. Chopek, S. Seidel, and J. Gergely, *Third International Biophysics Congress*, Cambridge U. P. (April 1969), p. 191.
253. M. M. Zaalishvili, *Physicochemical Principles of Muscular Activity* [in Russian], Metsniereba, Tbilisi (1971).
254. K. I. Laidler, *The Chemical Kinetics of Enzyme Action*, Oxford U. P. (1958).
255. V. A. Yakovlev, *Kinetics of Enzymatic Catalysis* [in Russian], Nauka, Moscow (1965).
256. T. Tokiwa, *Biochem. Biophys. Res. Commun.*, **44**, (2) (1971).
257. A. I. Kyaivyaryainen, R. S. Nezlin, M. V. Vol'kenshtein, and G. I. Likhtenshtein, *Fourth International Biophysics Congress, Section Reports* [in Russian], Vol. 4, Moscow (1972), p. 227.
258. L. J. Berliner and H. M. McConnell, *Biochem. Biophys. Res. Commun.*, **43**, 651 (1971).
259. J. C. Hsia, D. J. Kosman, and L. H. Piette, *Arch. Biochem. Biophys.*, **149**, 441 (1972).
260. D. J. Kosman and L. H. Piette, *Arch. Biochem. Biophys.*, **149**, 452 (1972).
261. H. L. Oppenheimer, B. Labouesse, and G. P. Hess, *J. Biol. Chem.*, **241**, 2720 (1966).
262. L. G. Ignat'eva and É. K. Ruuge, *Biofizika*, **17**, 148 (1972).
263. R. W. Burley, J. C. Seidel, and J. Gergely, *Arch. Biochem. Biophys.*, **150**, 792 (1972).
264. I. Tonomura, S. Watanabe, and M. Morales, *Biochemistry*, **8**, 2171 (1969).
265. T. Asacura, J. S. Leigh, H. R. Drott, T. Yonetani, and B. Chance, *Proc. Natl. Acad. Sci. U. S.*, **68**, 861 (1971).
266. J. S. Hsia and L. H. Piette, *Arch. Biochem. Biophys.*, **132**, 466 (1969).
267. T. V. Avilova, G. I. Likhtenshtein, and B. N. Vlasov, *Zh. Fiz. Khim.*, **46**, 281 (1972).
268. A. M. Vasserman, A. L. Buchachenko, A. A. Kovarskii, and M. B. Neiman, *Vysokomolekul. Soed.*, **AX**, 1930 (1968).
269. V. B. Stryukov and G. V. Korolev, *Vysokomolekul. Soed.*, **AXI**, 419 (1969).
270. A. L. Kovarskii, *Investigation of Molecular Dynamics and the Structure of Liquids and Polymers with Paramagnetic Probes* [in Russian], Master's Dissertation, Moscow (1972).
271. A. M. Wasserman, A. L. Buchachenko, A. L. Kovarskii, and M. B. Neiman, *Eur. Polymer J., Suppl.*, 473 (1969).
272. A. L. Kovarskii, A. M. Vasserman, and A. L. Buchachenko, Vysokomolekul. Soed., **AXIII**, 1647 (1971).
273. A. L. Kovarskii, A. M. Vasserman, and A. L. Buchachenko, *Bysokomolekul. Soed.*, **BXII**, 211 (1970).
274. A. L. Kovarskii, A. M. Vasserman, and A. L. Buchachenko, *Dokl. Akad. Nauk SSSR*, **193**, 132 (1970).
275. O. A. Zaporozhskaya, A. L. Kovarskii, V. S. Pudov, A. M. Vasserman, and A. L. Buchachenko, *Vysokomolekul. Soed.*, **BXII**, 702 (1970).
276. A. L. Kovarskii, S. G. Burkova, A. M. Vasserman, and Yu. L. Morozov, *Dokl. Akad. Nauk. SSSR*, **196**, 383 (1971).
277. Yu. I. Khurgin, Yu. M. Azizov, L. V. Abaturov, G. A. Kogan, and V. Ya. Roslyakov, *Biofizika*, **17**, (3) (1972).
278. A. L. Kovarskii, A. M. Vasserman, and A. L. Buchachenko, *Dokl. Akad. Nauk SSSR*, **201**, 121 (1971).
279. G. I. Likhtenshtein, E. Ya. Alfimova, L. A. Syrtsova, A. V. Kulikov, and T. I. Pisarskaya, *Eleventh European Congress on Molecular Spectroscopy, Theses* [in Russian] Tallin (1973), p. 37.
280. A. L. Buchachenko, A. M. Wasserman, and A. L. Kovarskii, *Internat. J. Chem. Kinet.*, **1**, 361 (1969).

281. C. Jolicoeur and H. L. Friedman, *J. Phys. Chem.*, **75**, 165 (1971).
282. A. Yu. Shaulov, A. M. Vasserman, and A. M. Brodskii, *Dokl. Akad. Nauk SSSR*, **196**, 888 (1971).
283. V. B. Stryukov, A. V. Dubovitskii, B. A. Rozenberg, and N. S. Enikolopov, *Dokl. Akad. Nauk SSSR*, **190**, 895 (1970).
284. V. B. Stryukov, É. G. Rozantsev, A. I. Kashlinskii, I. G. Mal'tseva, and I. F. Tabakov, *Dokl. Akad. Nauk SSSR*, **190**, 895 (1970).
285. I. G. Mal'tseva, V. B. Stryukov, and É. G. Rozantsev, *Dokl. Akad. Nauk SSSR*, **203**, 313 (1972).
286. L. I. Anziverova, N. N. Korst, W. B. Strukow, A. I. Ivanova, and I. B. Rabinkina, *Molec. Phys.*, **21** (1973).
287. V. B. Stryukov, A. M. Kraitsberg, and T. V. Sosnina, *Vysokomolekul. Soed.*, AXV, 1391 (1973).
288. E. N. Frolov, G. I. Likhtenshtein, F. Parek, and H. Formanic, *Studia Biophysica* (in press).
289. G. I. Likhtenshtein and T. V. Troshkina, *Molekul. Biol.*, **2**, 654 (1968).
290. T. V. Troshkina and G. I. Likhtenshtein, *Molekul. Biol.*, **2**, 659 (1968).
291. G. I. Likhtenshtein, in *The Role of Water in Biological Systems* [in Russian], Nauka, Moscow (1967).
292. G. I. Likhtenshtein, in *Water in Biological Systems*, Consultants Bureau, New York (1969).
293. G. I. Likhtenshtein and T. V. Avilova, *Usp. Sovrem. Biol.*, **75**, 26 (1973).
294. J. G. Calvert and J. N. Pitts, *Photochemistry*, Wiley, New York (1966).
295. S. V. Konev, *Electronic Excited States in Biopolymers* [in Russian], Nauka i Tekhnika, Minsk (1965).
296. Yu. A. Vladimirov, *Photochemistry of Luminescence of Proteins* [in Russian], Izd-vo AN SSSR (1961).
297. A. N. Terenin, *Photonics of Molecular Dyes and Related Organic Compounds* [in Russian], Nauka, Leningrad (1967).
298. É. A. Burshtein, *Itogi Nauki, Ser. Molekul Biol.*, **2** (1973).
299. S. M. Vavilov, *Microstructure of the World, Collected Works* [in Russian], Vol. II, Izd-vo AN SSSR, Moscow (1952).
300. T. Förster, *Disc. Faraday Soc.*, **27**, 7 (1959).
301. M. D. Galanin, *Trudy Fizicheskogo Instituta im. P. N. Lebedeva An SSSR*, **12**, 3 (1960).
302. G. Karreman, R. H. Steel, and A. Szent-Gyorgy, *Proc. Natl. Acad. Sci. U. S.*, **44**, 140 (1958).
303. A. P. Pivovarov, Yu. A. Ershov, and A. F. Lukovnikov, *Plastmassy*, **10**, 7 (1966).
304. E. Ya. Alfimova, L. A. Syrtsova, G. I. Likhtenshtein, and T. N. Pisarskaya, *Molekul. Biol,.* **8**, 676 (1974).
305. E. Ja. Alphimova, R. I. Gvosdev, and G. I. Likhtenshtein, *Studia Biophys.*, **44**, 93 (1974).
306. R. I. Gvozdev, A. P. Sadkov, E. Ya. Alfimova, A. I. Kotel'nikov, and G. I. Likhten, shtein, in *Mechanism of the Biological Fixation of Molecular Nitrogen* [in Russian]- Nauka, Moscow (1973).
307. C. Weber and F. I. Teal, *Disc. Faraday Soc.*, **27**, 134 (1959).
308. D. S. Markovich, M. V. Vol'kenshtein, and N. V. Umrikhin, *Molekul. Biol.*, **5**, 51 (1971).
309. V. I. Gol'danskii and R. Kherber, Editors, *Application of the Mössbauer Effect in Chemistry* [in Russian], Mir, Moscow (1970).

310. V. I. Gol'danskii, *The Mössbauer Effect and its Application in Chemistry* [in Russian]. Izd-vo AN SSSR, Moscow (1963).
311. G. K. Werthein, *Mössbauer Effect*, Academic, New York (1964).
312. The Mössbauer Effect, Wiley, New York (1962).
313. G. Frauenfelder, *The Mössbauer Effect*, Benjamin, New York (1962).
314. *Application of the Mössbauer Effect in Chemistry* [Russian translation], Mir, Moscow (1971).
315. E. F. Makarov, *Gamma-Resonance Spectroscopy of Chemical Compounds of Iron and Tin* [in Russian], Doctoral Dissertation, Moscow (1968).
316. G. V. Novikov, L. A. Syrtsova, V. A. Trukhtanov, and V. I. Gol'danskii, *Dokl. Akad. Nauk SSSR*, **196**, 390 (1971).
317. E. N. Frolov, G. V. Novikov, G. I. Likhtenshtein, V. A. Trukhtanov, and V. I. Gol'danskii, *Dokl. Akad. Nauk SSSR*, **196**, 390 (1971).
318. V. A. Trukhtanov, *The Mössbauer Effect in Solids* [in Russian], Master's Dissertation, Moscow (1965).
319. P. P. Craig and N. Sutin, *Phys. Rev. Lett.*, **11**, 460 (1963).
320. D. C. Champeney, *Phys. Bull.*, **21**, 248 (1970).
321. *Chemical Applications of Mössbauer Spectroscopy* [Russian translation] Mir, Moscow (1970).
322. F. Parek and H. Formanic, *Acta Cryst.*, **A-27**, 573 (1971).
323. Ts. Bonchev, P. Aidemirski, I. Mandzhukov, N. Medelyakova, B. Skorchev, and A. Strigachev, *Zh. Éksperim. i Teor. Fiz.*, **50**, 62 (1966).
324. N. A. Kiselev, *Electron Microscopy of Biological Molecules* [in Russian], Nauka, Moscow (1965).
325. R. I. Gvozdev, A. P. Sadkov, L. A. Levchenko, and A. V. Kulikov, *Izv. Akad. Nauk SSSR, Ser. Biol.*, No. 2, 246 (1971).
326. G. I. Slepko, A. M. Uzenskaya, V. R. Linde, and L. A. Levchenko, *Izv. Akad. Nauk SSSR, Ser. Biol.*, No. 1, 86 (1971).
327. G. I. Likhtenshtein, in *Mechanism of the Biological Fixation of Molecular Nitrogen* [in Russian], Nauka, Moscow (1973).
328. L. A. Levchenko, A. V. Raevskii, R. I. Gvozdev, A. P. Sadkov, and I. Z. Mitsova, *Izv. Akad. Nauk SSSR, Ser. Biol.*, 140 (1975).
329. L. G. Makarova and A. N. Nesmeyanov, *Methods of Investigating Organometallic Compounds, Mercury* [in Russian], Nauka, Moscow (1965).
330. G. G. Hammes and P. Fasella, *J. Amer. Chem. Soc.*, **85**, 3929 (1963).
331. M. Dixon and E. Webb, *Enzymes*, Academic, New York (1964).
332. *Classification and Nomenclature of Enzymes* [Russian translation], IL, Moscow (1962).
333. A. E. Braunshtein, Editor, Enzymes [in Russian], Nauka, Moscow (1964).
334. J. R. Koshland and K. E. Neet, *Ann. Rev. Biochem.*, **37**, 359 (1968).
335. F. H. Johnson, H. Eyring, and M. I. Polissar, *The Kinetic Basis of Molecular Biology*, New York (1954).
336. V. L. Strayer, *Ann. Rev. Biochem.*, **37**, 25 (1968).
337. C. C. Blave, L. N. Iohnson, G. A. Mair, A. C. Notrth, D. C. Phillips, and V. R. Sarma, *Proc. Roy. Soc.*, **B167**, 378 (1967).
338. G. I. Likhtenshtein, *Biofizika*, **11**, 24 (1966).
339. R. Lumry and H. I. Eyring, *J. Phys. Chem.*, **58**, 110 (1954).
340. K. Linderström-Lang and J. Shelman, *The Enzymes*, **1**, 443 (1959).
341. D. R. Strom and D. E. Koshland, *Proc. Natl. Acad. Sci. U. S.* **66**, 445 (1970).
342. M. V. Vol'kenshtein, in *Molecular Biology* [in Russian], Nauka, Moscow (1966).

343. G. G. Hammes, *J. Amer. Chem. Soc.*, **88**, 5607 (1966).
344. G. G. Hammes, *Acc. Chem. Res.*, **1**, 321 (1968).
345. V. I. Ivanov, M. Ya. Karpeiskii, and A. E. Braunshtein, in *Pyridoxal Catalysis* [in Russian], Mir, Moscow (1968).
346. R. Lumry and S. Rajender, *Biopolymers*, **9**, 1125 (1970).
347. V. I. Ivanov and M. Ya. Karpeisky, *Adv. Enzymol.*, **32**, 21 (1969).
348. G. G. Hammes, *Seventh International Symposium on the Chemistry of Natural Compounds, Symposium 1* [in Russian], Riga (1970).
349. Yu. I. Khurgin, D. S. Chernyavskii, and S. É. Shnol', *Molekul. Biol.*, **1**, 419 (1967).
350. N. O. Kobozev, *Usp. Khim.*, **25**, 545 (1956).
351. S. Bernhard, *The Structure and Function of Enzymes*, Benjamin, New York (1968).
352. O. M. Poltorak and E. S. Chukhrii, *Physicochemical Basis of Enzymatic Catalysis* [in Russian], Vysshaya Shkola, Moscow (1971).
353. W. Jencks, *Catalysis in Chemistry and Enzymology* [Russian translation], edited by V. V. Berezin of the USSR Academy of Sciences, Mir, Moscow (1972).
354. W. P. Jencks, *Catalysis in Chemistry and Enzymology*, McGraw-Hill, New York (1969).
355. M. V. Vol'kenshtein, *Molecules and Life* [in Russian], Nauka, Moscow (1965).
356. M. V. Vol'kenshtein, *The Physics of Enzymes* [in Russian], Nauka, Moscow (1967).
357. M. V. Vol'kenshtein, *Abstracts of the Seventh Meeting of FEBS*, Varna (1971), p. 109.
358. M. V. Volcenstein, *J. Theor. Biol.*, **34**, 193 (1972).
359. M. V. Vol'kenshtein, R. D. Dogonadze, A. H. Madumarov, Z. D. Urushadze, and Yu. I. Kharkats, *Molekul. Biol.*, **6**, 341 (1972).
360. R. Biltonen and R. Lumry, *Structure and Stability of Biology Macromolecules*, Dekkar, New York (1969).
361. L. Perutz, *Nature*, **228**, 726 (1970).
361a. L. A. Blyumenfel'd and V. K. Kol'tover, *Molekul. Biol.*, **6**, 161 (1972).
362. D. H. Meadows and O. Jardetsky, *Proc Natl. Acad. Sci. U. S.*, **61**, 4061 (1968).
363. L. A. Blyumenfel'd, *Biofizika*, **16**, 724 (1971).
364. L. A. Blyumenfel'd, *Biofizika*, **17**, 584 (1972).
365. G. I. Likhtenshtein, Yu. B. Grebenshikov, T. B. Avilova, E. N. Frolov, and Yu. D. Achmedov, *Abstracts of tne Seventh Meeting of FEBS*, Varna (1971), p. 109.
366. K. Martinek and I. V. Berezin, *Zh. Vsesoyuzn. Khim. Ob-va*, **16**, 362 (1971).
367. V. A. Yakovlev, *Zh. Vsesoyuzn. Khim. Ob-va*, **16**, 391 (1971).
368. S. E. Bresler, *Introduction to Molecular Biology* [in Russian], Nauka, Moscow (1966).
369. T. C. Bruice and S. J. Benkovic, *Biorganic Mechanisms*, Benjamin, New York (1966).
370. V. Ya. Roslyakov and Yu. I. Khurgin, *Izv. Akad. SSSR, Ser. Khim.*, **1366** (1971).
371. K. Martinek, *Kinetic and Thermodynamic Laws of Enzymatic Catalysis* [in Russian], Doctoral Dissertation, Moscow (1972).
372. V. N. Dorovska, S. D. Varfolomeev, and I. V. Berezin, *Biochim. Biophys. Acta.* **271**, 80 (1972).
373. V. N. Kondrat'ev, *Kinetics of Gas-Phase Chemical Reactions* [in Russian], Izd-vo Akad. Nauk SSSR (1958).
374. S. Benson, Thermochemical Kinetics, Wiley, New York (1968).
375. E. A. Moelwyn-Hughes, *Physical Chemistry*, Pergamon, New York (1961).
376. L. E. Shoke and H. Neurath, *J. Biochem.*, **182**, 577 (1950).
377. Y. B. Wilson and E. Cabib, *J. Amer. Chem. Soc.*, **78**, 202 (1956).
378. M. Monnot, *Biochim. Biophys. Acta*, **93**, 31 (1964).
379. R. J. Seguin and G. W. Kosieki, *Canad. J. Biochem.*, **45**, 659 (1967).
380. U. Massey, *Biochem. J.* **53**, 72 (1953).

381. G. I. Slepko, *Investigation of Nitrogenase* [in Russian], Master's Dissertation, Moscow (1971).
382. N. Azuma and Y. Tonomura, *Biochim. Biophys. Acta*, **73**, 499 (1963).
383. B. D. Sykes, *Biochemistry*, **8**, 1110 (1969).
384. B. A. Wittenberg, E. Antonioni, M. Brunori, R. W. Noble, J. B. Wittenberg, and E. V. Wyman, *Biochemistry*, **6**, 1970 (1967).
385. B. Belleau and J. L. Lavoie, *Canad. J. Biochem.*, **46**, 1397 (1968).
386. H. Kaplan and K. J. Laidler, *Canad. J. Biochem.*, **45**, 547 (1967).
387. G. G. Hammes and J. F. Tancredi, *Biochim. Biophys. Acta*, **146**, 312 (1967).
388. A. Schejter and I. Aviram, *Biochemistry*, **8**, 149 (1969).
389. K. J. Laidler, *Arch. Biochem. Biophys.*, **44**, 338 (1953).
390. J. Tonomura, S. Tokura, and K. Sekija, *J. Biol. Chem.*, **237**, 1074 (1962).
391. R. L. Nath and I. Das, *Enzymologia*, **26**, 269 (1963).
392. R. S. Burns, *Biochim. Biophys. Acta*, **171**, 253 (1969).
393. L. M. Raikhman, V. S. Belova, M. R. Borukaeva, and G. I. Likhtenshtein, *Biokhimiya*, **36**, 674 (1971).
394. J. L. Webb, *Enzyme and Metabolic Inhibitors*, Academic, New York (1966).
395. G. Nemethy and H. A. Scheraga, *J. Chem. Phys.*, **36**. 3401 (1962); *Biopolymers*, **1**, 43 (1963).
396. D. G. Doherty and E. Vaslow, *J. Amer. Chem. Soc.*, **74**, 931 (1952).
397. N. A. Gabelova and A. A. Zamyatnin, *Studia Biophysica*, **20**, 35 (1970).
398. A. A. Zamyatin, *Zh. Fiz. Khim.*, **45**, 1007 (1971).
399. A. J. Hymes, C. C. Cuppet, and W. J. Canady, *J. Biol. Chem.*, **244**, 637 (1969).
400. G. G. Hammes and P. R. Schimmel, *J. Amer. Chem. Soc.*, **87**, 4665 (1965).
401. J. Heidberg, E. Holler, and H. Hartmann, *Ber. Bensenges. Phys. Chem.*, **71**, 19 (1967).
402. G. I. Likhtenshtein and B. I. Sukhorukov, *Zh. Fiz. Khim.*, **38**, 747 (1964).
403. W. R. Guild and R. P. van Tabergen, *Science*, **125**, 939 (1957).
404. M. V. Vol'kenshtein, *Configurational Statistics of Polymer Chains* [in Russian], Izd-vo AN SSSR, Moscow (1959).
405. B. H. Zimm and J. K. Bragg, *J. Chem. Phys.*, **31**, 526 (1959).
406. M. V. Vol'kenshtein, *Biofizika*, **6**, 257 (1961).
407. O. B. Ptitsyn and T. M. Birshtein, *Conformations of Macromolecules* [in Russian], Nauka, Moscow (1964).
408. O. B. Ptitsyn, *Usp. Biol. Nauk*, **69**, 26 (1970).
409. H. Eyring and A. Stearn, *Chem. Rev.*, **24**, 253 (1939).
410. M. V. Vol'kenshtein, Yu. L. Gotlib, and O. B. Ptitsin, *Fiz. Tverd. Tela*, **3**, 420 (1961).
411. Yu. L. Gotlib, *Fiz. Tverd. Tela*, **3**, 2170 (1961).
412. D. E. Koshland, G. Nemethy, and D. Filmer, *Biochemistry*, **5**, 365 (1966).
413. M. V. Vol'kenshtein, in *Molecular Biophysics* [in Russian], Nauka, Moscow (1965).
414. B. I. Sukhorukov and G. I. Likhtenshtein, *Biofizika*, **10**, 935 (1965).
415. G. I. Likhtenshtein and B. I. Sukhorukov, *Zh. Fiz. Khim.*, **38**, 747 (1964).
416. G. I. Likhtenshtein, *Zh. Fiz. Khim.*, **44**, 1913 (1970).
417. J. S. Leffler, *J. Org. Chem.*, **20**, 1202 (1955).
418. A. Hvidt and O. S. Nielsen, *Adv. Protein Chem.*, **21**, 288 (1966).
419. L. V. Abaturov, F. V. Shmakova, and Ya. M. Varshavskii, *Molekul. Biol.*, **4**, 831 (1970).
420. S. N. Zelenin and M. L. Khidekel', *Usp. Khim.*, **39**, 209 (1971).
421. A. A. Balandin, *The Contemporary State of the Multiplet Theory of Heterogeneous Catalysis* [in Russian], Nauka, Moscow (1968).
422. N. N. Semenov, *Chain Reactions* [in Russian], Izd-vo AN SSSR, Moscow (1934).

423. I. N. Bronsted and K. Z. Pedersen, *J. Phys. Chem.*, **108,** 185 (1924).
424. R. A. Marcus, *J. Chem. Phys.*, **26,** 872 (1957).
425. G. G. Swain, *J. Amer. Chem. Soc.*, **70,** 119 (1948).
426. G. N. Lewis, *Valence and the Structure of Atoms and Molecules*, Chem. Catalog Co., New York (1923), p. 113.
427. F. London, *Z. Electrochem.*, **35,** 552 (1925).
428. Zh. P. Kachanova, Yu. M. Azizov, and A. P. Purmal', in *Problems in Kinetics and Catalysis. XIII. Complexation in Catalysis* [in Russian], Nauka, Moscow (1968).
429. Yu. A. Zhdanov and V. I. Minkin, *Correlation Analysis in Organic Chemistry* [in Russian], Izd. Rostovskogo Un-Ta (1966).
430. V. A. Palm, *Fundamentals of the Quantitative Theory of Organic Reactions* [Russian translation], Khimiya, Leningrad (1967).
431. K. I. Matveev, *Kinetika i Kataliz*, **13,** 874 (1972).
432. G. S. Hammond, *J. Amer. Chem. Soc.*, **77,** 334 (1955).
433. T. G. Bordwell, *Acc. Chem. Res.*, **3,** 281 (1970).
434. J. Hine, *J. Org. Chem.*, **31,** 1236 (1966).
435. F. O. Rice and E. Teller, *J. Chem. Phys.*, **6,** 489 (1938).
436. R. Woodward and R. Hoffman, *The Conservation of Orbital Symmetry* [Russian translation], Mir, Moscow (1971).
437. R. Woodward and R. Hoffman, *The Conservation of Orbital Symmetry* Weinheim (1970).
438. G. B. Gill, *Usp. Khim.*, **XL,** 1105 (1971) [Russian translation].
439. G. B. Gill, *Chem. Soc. Quart. Rev.*, **22,** 228 (1968).
440. K. Trindl, *J. Amer. Chem. Soc.*, **92,** 3251 (1970).
441. K. Trindl, *Usp. Khim.*, **XL,** 2102 (1971) [Russian Translation].
442. P. Nichols, *The Enzymes*, **8,** 147 (1963).
443. S. Rajender, R. Lumry, and M. Han., *J. Chem. Phys.*, **75,** 1375 (1971).
444. K. Yagi, M. Nishikimi, N. Ohishi, and K. Hiromi, *J. Biochem.*, **65,** 663 (1969).
445. K. Yagi, M. Nishikimi, N. Ohishi, and A. Takai, *FEBS Lett.*, **6,** 22 (1970).
446. K. Yagi and N. Nishikimi, *Biochem. Biophys. Res. Commun.*, **34,** 549 (1969).
447. R. D. Kornberg and H. M. McConnell, *Proc. Natl. Acad. Sci. U. S.*, **68,** 2564 (1971).
448. H. M. McConnell, P. Devaux, and C. Scandella, in *Membrane Research*, Academic, New York–London (1972).
449. S. Schreier-Mucillo, D. Marsh, H. Dugas, H. Schneider, and I. C. P. Smith, *Chem. Phys. Lipids*, **10,** 11 (1973).
450. W. L. Hubbell and H. M. McConnell, *Proc. Natl. Acad. Sci. U. S.*, **61,** 12 (1968).
451. A. D. Keith, A. S. Waggoner, and O. H. Griffith, *Proc. Natl. Acad. Sci. U. S.*, **61,** 819 (1968).
452. B. McFarland, in Fleischer, Packer, and Estabrook, Editors, *Methods in Enzymology*, Academic, New York–London (1973).
453. W. Hubbell and H. M. McConnell, *Proc. Natl. Acad. Sci. U. S.*, **63,** 16 (1969).
454. S. Ohnishi and H. M. McConnell, *J. Amer. Chem. Soc.*, **87,** 2293 (1965).
455. A. Saupe, *Z. Naturforsch.*, **19,** 161 (1964).
456. J. Seelig, *J. Amer. Chem. Soc.*, **92,** 3881 (1970).
457. M. R. Hemmings and H. J. C. Berendsen, *J. Magnet. Resonance*, **8,** 133 (1972).
458. A. N. Kuznetsov, V. A. Livshits, G. G. Malenkov, L. A. Mel'nik, and B. G. Tenchov, *Proceedings of the Second Scientific Conference on Liquid Crystals* [in Russian], Ivanovo (1973).
459. J. C. Williams, R. Mehlhorn, and A. D. Keith, *Chem. Phys. Lipids*, **7,** 207 (1971).

460. T. Kawamura, S. Matsunami, and I. Yonesawa, *Bull. Chem. Soc. Japan*, **40,** 1111 (1967).

461. V. K. Kol'tover, *Investigation of Electron-Transporting Membranes with the Aid of Molecular Probes* [in Russian], Master's Dissertation, Moscow (1971).

462. V. K. Kol'tover, *Biofizika*, **18,** 661 (1973).

463. B. Annaev, V. K. Kol'tover, L. M. Raikhman, and V. I. Suskina, *Dokl. Akad. Nauk SSSR*, **196,** 969 (1971).

464. V. K. Koltover, L. M. Reichman, A. A. Jasaitis, and L. A. Blumenfeld, *Biochim. Biophys. Acta*, **234,** 306 (1971).

465. V. K. Kol'tover, L. M. Raikhman, A. A. Yasaitis, and L. A. Blyumenfel'd, *Dokl. Akad. Nauk SSSR*, **197,** 219 (1971).

466. V. K. Kol'tover, L. M. Raikhman, A. A. Yasaitis, and L. A. Blyumenfel'd, *Fourth International Biophysics Congress, Section Reports* [in Russian], Vol. 1, Moscow (1972), p. 77.

467. A. S. Waggoner, O. H. Griffith, and C. R. Christensen, *Proc. Natl. Acad. Sci. U. S.*, **57,** 1198 (1967).

468. A. S. Waggoner, A. D. Keith, and O. H. Griffith, *J. Phys. Chem.*, **72,** 4129 (1968).

469. M. G. Goldfield, V. K. Koltover, and E. G. Rozantzev, *Kolloid-Z. und Z. Polymere*, **243,** 62 (1971).

470. Y.-K. Levine, P. Partington, G. C. K. Roberts, N. J. M. Birdsall, A. G. Lee, and J. C. Metcalfe, *FEBS Lett.*, **23,** 203 (1972).

471. A. S. Waggoner, T. J. Kingzett, S. Kottschaefer, O. H. Griffith, and A. D. Keith, *Chem. Phys. Lipids*, **3,** 245 (1969).

472. E. Oldfield and D. Chapman, *Biochem. Biophys. Res. Commun.*, **43,** 610 (1971).

473. J. S. Hsia, H. Schneider, and I. C. P. Smith, *Chem. Phys. Lipids*, **4,** 238 (1970).

474. D. Chapman, D. F. H. Wallach, in D. D. Chapman, Editor, *Biological Membranes*, Academic, New York–London (1968).

475. K. Hong and W. L. Hubbell, *Proc. Natl. Acad. Sci. U. S.*, **69,** 2617 (1972).

476. R. D. Lapper, S. J. Paterson, and I. C. P. Smith, *Canad. J. Biochem.*, **50,** 969 (1972).

477. P. Jost, J. L. Libertini, V. Hebert, and O. H. Griffith, *J. Molec. Biol.*, **59,** 77 (1971).

478. B. G. McFarland and H. M. McConnell, *Proc. Natl. Acad. Sci. U. S.*, **68,** 1271 (1971).

479. K. W. Butler, H. Dugas, I. C. P. Smith, and Schneider, *Biochem. Biophys. Res. Commun.*, **40,** 770 (1970).

480. S. J. Paterson, K. W. Butler, P. Huang, J. Labelle, I. C. P. Smith, and H. Schneider, *Biochim. Biophys. Acta*, **266,** 597 (1972).

481. S. P. Verma, H. Schneider, and I. C. P. Smith, *FEBS Lett.*, **25,** 197 (1972).

482. Y. K. Levine, in *Progress in Biophysics and Molecular Biology*, Vol. 24, Oxford (1972), pp. 1–74.

483. H. Träuble, *Naturwissenschaften*, **58,** 277 (1971).

484. M. D. Barratt, D. K. Green, and D. Chapman, *Chem. Phys. Lipids*, **3,** 140 (1969).

485. E. Sackmann and H. Träuble, *J. Amer. Chem. Soc.*, **94,** 4482, 4492 (1972).

486. H. Träuble and E. Sackmann, *J. Amer. Chem. Soc.*, **94,** 4499 (1972).

487. R. A. Long, F. E. Hruska, and H. D. Geesser, *Biochem. Biophys. Res. Commun.*, **45,** 167 (1971).

488. P. Devaux and H. M. McConnell, *J. Amer. Chem. Soc.*, **94,** 4475 (1972).

489. R. D. Kornberg and H. M. McConnell, *Biochemistry*, **10,** 1111 (1971).

490. M. D. Barratt, D. K. Green, and D. Chapman, *Biochim. Biophys. Acta*, **152,** 20 (1968).

491. K. U. Berger, M. D. Barratt, V. B. Kamat, *Biochem. Biophys. Res. Commun.*, **40,** 1273 (1970).

492. M. D. Barratt, R. B. Leslie, and A. M. Scanu, *Chem. Phys. Lipids*, **7**, 345 (1971).
493. M. D. Barratt and L. Rayner, *Biochim. Biophys. Acta*, **255**, 974 (1972).
494. A. M. Gotto, H. Kohn, M. E. Birnbaumer, *Proc. Natl. Acad. Sci. U. S.*, **65**, 145 (1970).
495. S. K. Devi, J. D. Morrisett, P. L. Jackson, and A. M. Gotto, *Fourth International Biophysics Congress, Section Reports* [in Russian], Vol. 1 Moscow (1972), p. 78.
496. S. Rottem, W. L. Hubbell, L. Hayflick, and H. M. McConnell, *Biochim. Biophys. Acta*, **219**, 104 (1970).
497. K. Mukai, C. M. Lang, and D. B. Chesnut, *Chem. Phys. Lipids*, **9**, 196 (1972).
498. W. L. Hubbell and H. M. McConnell, *Proc. Natl. Acad. Sci. U. S.*, **64**, 20 (1969).
499. F. R. Landsberger, J. Lenard, J. Paxton, and R. W. Compans, *Proc. Natl. Acad. Sci. U. S.*, **68**, 2579 (1971).
500. M. Nakamura and S. Ohnishi, *Biochem. Biophys. Res. Commun.*, **46**, 926 (1972).
501. M. E. Tourtellotte, E. D. Branton, and A. D. Keith, *Proc. Natl. Sci. U. S.*, **66**, 904 (1970).
502. N. Z. Stanacev, L. Stuhno-Sekalec, S. L. Schreier-Mucillos, and I. C. P. Smith, *Biochem. Biophys. Res. Commun.*, **46**, 114 (1972).
503. A. Colleau, P. M. Vignais, and L. H. Piette, *Biochem. Biophys. Res. Commun.*, **45**, 1495 (1972).
504. J. Finean, *Quart. Rev. Biophys.*, **2**, 1 (1969).
505. S. Koga, A. Eshigo, and K. Nunomura, *Biophys. J.*, **6**, 665 (1968).
506. Yu. A. Kutlakhmedov, V. K. Kol'tover, and Yu. V. Kokhanov, in *The Effect of Ionizing Radiation on Cell Membranes* [in Russian], Atomizdat, Moscow (1973).
507. H. Simpkins, E. Panko, and Tay-Sin, *Canad. J. Biochem.*, **50**, 174 (1972).
508. W. L. Hubbell, J. C. Metcalfe, and H. M. McConnell, *Biochim. Biophys. Acta*, **219**, 415 (1970).
509. J. Kroes, R. Ostwald, and A. D. Keith, *Biochim. Biophys. Acta*, **71**, 274 (1972).
510. V. P. Skulachev, *Accumulation of Energy in the Cell* [in Russian], Nauka, Moscow (1969).
511. G. Zimmer, L. Packer, R. Mehlhorn, and A. D. Keith, *Hoppe-Seylers Z. physiol. Chem.*, **353**, 1578 (1972).
512. B. Annaev, V. K. Kol'tover, O. N. Mamedniyazov, L. M. Raikhman, and É. G. Rozantsev, *Biofizika*, **17**, 224 (1972).
513. H. M. McConnell, L. K. Wright, and B. G. McFarland, *Biochem. Biophys. Res. Commun.*, **47**, 273 (1972).
514. H. E. Sandberg, R. G. Bryant, and L. H. Piette, *Arch. Biochem. Biophys.*, **133**, 144 (1969).
515. D. Chapman, M. D. Barratt, and V. B. Kamat, *Biochim. Biophys. Acta*, **173**, 154 (1969).
516. D. D. Johnes, A. Haug, M. Jost, and D. R. Graber, *Arch. Biochem. Biophys.*, **135**, 296 (1969).
517. R. Cooke and M. F. Morales, *Biochemistry*, **8**, 3188 (1969).
518. J. D. Morrisett and H. R. Drott, *J. Biol. Chem.*, **244**, 5083 (1969).
519. W. C. Landraf and G. Inesi, *Arch. Biochem. Biophys.*, **130**, 111 (1969).
520. G. Inesi and W. C. Landraf, *J. Bioenerg.*, **1**, 355 (1970).
521. M. Nakamura, H. Hori, and T. Mitsui, *J. Biochem.*, **72**, 635 (1972).
522. L. M. Pozharitskaya, V. K. Kol'tover, N. S. Agre, and L. V. Kalakutskii, *Mikrobiologiya*, **40**, 1110 (1971).
523. E. A. Lapteva, L. M. Pozharitskaya, E. N. Frolov, V. K. Kol'tover, N. S. Agre, and L. V. Kalakutskii, *Izv. Akad. Nauk SSSR, Ser. Biol.*, No. 6 (1973).

524. A. Azzi, A. M. Tamburo, G. Farnia, and E. Cobbe, *Biochim. Biophys. Acta*, **256,** 619 (1972).

525. V. I. Binyukov, S. F. Borunova, M. G. Gol'dfel'd, I. G. Zhukova, D. G. Kudlai, A. N. Kuznetsov, A. B. Shapiro, and D. N. Ostrovskii, *Biokhimiya*, **36,** 1149 (1971).

526. V. I. Binyukov, A. S. Kaprel'yants, A. N. Kuznetsov, and D. N. Ostrovskii, *Fourth International Biophysics Congress, Section Reports* [in Russian], Vol. 3, Moscow (1972), p. 16.

527. L. M. Pozharitskaya, L. M. Raikhman, L. V. Kalakutskii, and V. K. Kol'tover, *Mikrobiologiya*, **41,** 175 (1972).

528. V. K. Kol'tover and L. A. Blyumenfel'd, *Fourth International Biophysics Congress, Section Reports* [in Russian), Vol. 3, Moscow (1972), p. 15.

529. V. K. Kol'tover and L. A. Blyumenfel'd, *Biofizika*, **18** (1973).

530. J. K. Raison, J. M. Lyons, R. J. Mehlhorn, and A. D. Keith, *J. Biol. Chem.*, **246,** 4036 (1971).

531. A. D. Keith and R. J. Mehlhorn, *Chem. Phys. Lipids*, **8,** 316 (1972).

532. A. Kemp, G. Groot, and A. Reitsma, *Biochim. Biophys. Acta*, **180,** 23 (1969).

533. J. F. Blazyk and J. M. Steim, *Biochim. Biophys. Acta*, **266,** 737 (1972).

534. M. Estafahani, A. R. Limbrick, S. Knutton, T. Oka, and S. J. Wakil, *Proc. Natl. Acad. Sci. U. S.*, **68,** 3180 (1971).

535. P. Overath, H.-U. Schaier, F.-F. Hill, and I. Lamneck-Hirsch, *Dynamic Structure of Cell Membranes. 22nd Colloq. Ges. Biol. Chem.* 15–17 *April 1971*, Mosbach/Baden— Berlin (1971), p. 149.

536. R. E. Barnett and C. M. Grisham, *Biochem. Biophys. Res. Commun.*, **48,** 1362 (1972).

537. M. Eigen and G. G. Hammes, *Adv. Enzymol.*, **25,** 1 (1963).

538. J. L. Rigaud, Y. Lange, E. M. Gary-Bobo, A. Samson, and M. Ptak, *Biochem. Biophys. Res. Commun.*, **50,** 59 (1973).

539. N. G. Taganov, M. G. Gol'fel'd, R. M. Davydov, and É. G. Rozantsev, *Ah. Fiz. Khim.*, **44,** 2402 (1970).

540. E. Weaver and H. P. Chon, *Science*, **153,** 301 (1966).

541. G. A. Corker, M.-P. Klein, and M. Calvin, *Proc. Natl. Acad. Sci. U. S.*, **56,** 1365 (1966).

542. N. V. Karapetyan, V. K. Kol'tover, I. N. Krakhmalaeva, and A. A. Krasnovskii, *Biofizika*, **16,** 1138 (1971).

543. V. S. Marinov, K. M. L'vov, B. I. Sukhorukov, L. P. Kayushin, and G. V. Postnikova, *Biofizika*, **16,** 337 (1971).

544. V. S. Marinov, M. R. Tairbekov, and Z. P. Gribova, *Fourth International Biophysics Congress, Section Reports* [in Russian], Vol. 1, Moscow (1972), p. 336.

545. L. S. Yaguzhinskii, V. M. Chumakov, V. P. Ivanov, V. V. Chistyakov, É. G. Rozantsev, and A. É. Kalmanson, *Dokl. Akad. Nauk SSSR*, **197,** 969 (1971).

546. A. I. Tsapin, Ju. G. Molotcovscy, M. G. Goldfeld and B. S. Dzjubenko, *Eur. J. Biochem.*, **20,** 218 (1971).

547. A. G. Chetverikov, L. A. Blyumenfel'd, and G. V. Fomin, *Biofizika*, **10,** 477 (1965).

548. A. G. Chetverikov, D. N. Kafalieva, and A. F. Vanin, *Biofizika*, **14,** 8 (1969).

549. V. M. Chumakov, G. L. Grigoryan, É. G. Rozantsev, and A. É. Kalmanson, *Biofizika*, **16,** 564 (1972).

550. A. Szent-Gyoergyi, *Horizons in Biochemistry*, Academic, New York, (1962).

551. S. S. Voronina and G. L. Grigoryan, *Izv. Akad. Nauk SSSR, Ser. Biol.*, No. 5, 723 (1972).

552. P. Emmerson and P. Howard-Flanders, *Radiat. Res.*, **26,** 54 (1965).

553. T. Brustad, *Internat. J. Rad. Biol.*, **22,** 443 (1972).

554. N. P. Konovalova, G. N. Bogdanov, V. B. Miller, M. B. Neiman, É. G. Rozantsev, and N. M. Émmanuél', *Dokl. Akad. Nauk. SSSR*, **157**, 707 (1964).

555. J. C. Hsia, W. L. Chen, L. T. Wong, R. A. Long, and W. Kalow, *Biochem. Biophys. Res. Commun.*, **48**, 1273 (1972).

556. P. Shofield, B. H. Hoffman, and A. Rich, *Biochemistry*, **9**, 2525 (1970).

557. B. H. Hoffman, P. Shofield, and A. Rich, *Proc. Natl. Acad. Sci. U. S.*, **62**, 1195 (1969).

558. I. C. P. Smith and P. Jamane, *Proc. Natl. Acad. Sci. U. S.*, **58**, 884 (1967).

559. H. Hara, T. Horiuchi, N. Saneyoshi, and S. Nishimura, *Biochem. Biophys. Res. Commun.*, **38**, 305 (1970).

560. R. I. Artyukh, G. B. Postnikova, B. I. Sukhorukov, and S. G. Kamzolova, *Biokhimiya*, **37**, 902 (1972).

561. V. P. Kumarev and D. G. Knorre, *Dokl. Akad. Nauk SSSR*, **193**, 103 (1970).

562. A. S. Girshouich, M. A. Grachev, D. G. Knorre, V. P. Kumarev, and V. J. Levintal, *FEBS Lett.*, **14**, 199 (1971).

563. V. I. Levental', Zh. M. Bekker, Yu. N. Molin, V. P. Kumarev, M. A. Grachev, and D. G. Knorre, *FEBS Lett.*, **24**, 149 (1972).

564. G. N. Bogdanov, V. N. Varfolomeev, and V. M. Shmonina, *Biofizika*, **19** (1974).

565. A. B. Shapiro, V. I. Suskina, B. V. Rozynov, and É. G. Rozantsev, *Izv. Akad. Nauk SSSR, Ser. Khim.*, No. 12, 2828 (1969).

566. D. Kabat, B. M. Hoffman, and A. Rich, *Biopolymers*, **9**, 95 (1970).

567. S. G. Kamzolova, A. I. Kolontarov, A. I. Petrov, L. I. Elfimova, and B. I. Sukhorukov, *Fourth International Biophysics Congress, Section Reports* [in Russian], Vol. 2, Moscow (1972), p. 264.

568. A. I. Petrov, G. B. Postnikova, and B. I. Sukhorukov, *Izv. Akad. Nauk SSSR, Ser. Khim.*, No. 6, 1453 (1972).

569. S. G. Kamzolova, A. I. Kolontarov, L. I. Elfimova, and B. I. Sukhorukov, *Dokl. Akad. Nauk SSSR*, **208**, 245 (1973).

570. E. M. Mil', G. L. Grigoryan, and K. E. Kruglyakova, *Fourth International Biophysics Congress, Section Reports* [in Russian], Vol. 2, Moscow (1972), p. 234.

571. B. I. Sukhorukov, A. M. Vasserman, L. I. Kozlova, and A. L. Buchachenko, *Dokl. Akad. SSSR*, **177**, 454 (1967).

572. A. A. Kraevskii, P. P. Purygin, L. N. Rudzite, Z. S. Belova, and B. P. Gottikh, *Izv. Akad. Nauk SSSR, Ser. Khim.*, No. 2, 378 (1968).

573. B. I. Sukhorukov and L. A. Kozlova, *Biofizika*, **15**, 539 (1970).

574. P. L. Privalov, G. M. Mrevlishvili, *Biofizika*, **12**, 22 (1967).

575. W. C. J. Ross, Biological Alkylating Agents, Butterworth, London (1962).

576. N. K. Kochetkov, É. I. Budovskii, E. D. Sverdlov, N. A. Simukova, N. F. Torchinskii, and V. N. Shibaev, *Organic Chemistry of Nucleic Acids* [in Russian], Khimiya, Moscow (1970).

577. D. P. Iton and B. D. Phillips, *Advances in Magnetic Resonance*, Vol. 1, Academic, New York (1965).

578. R. J. Kurland, *J. Magnet. Resonance*, **2**, 281 (1970).

579. E. D. Becker, *Appl. Spectroscopy*, **26**, 421 (1972).

580. O. B. Ptitsyn, *Fourth International Biophysics Congress, Symposium Reports* [in Russian], Moscow (1972), p. 8.

581. V. I. Lim, *Fourth International Biophysics Congress, Section Reports* [in Russian], Vol. 2, Moscow (1972), p. 29.

582. A. V. Finkel'shtein and O. B. Ptitsyn, *Fourth International Biophysics Congress, Section Reports* [in Russian], Vol. 2, Moscow (1972), p. 27.

583. S. A. Kozitsyn, V. I. Lim, and O. B. Ptitsyn, *Fourth International Biophysics Congress Section Reports* [in Russian], Vol. 2, Moscow (1972), p. 32.

Translated Russian Journals

Biokhimiya—Biochemistry (USSR)
Dokl. Akad. Nauk—Doklady
Fiziol. Rastenii—Soviet Plant Physiology
Fiz. Tverd. Tela—Soviet Physics–Solid State
Izv. Akad. Nauk SSSR, Ser. Khim.—Bulletin of the Academy of Sciences of the USSR,
   Division of Chemical Science
Khim. Geterotsikl. Soedin.—Chemistry of Heterocyclic Compounds
Khim. Prirodn. Soedin.—Chemistry of Natural Compounds
Kinetika i Kataliz—Kinetics and Catalysis
Mikrobiologiya—Microbiology
Molekul. Biol.—Molecular Biology
Opt. i Spektrosk.—Optics and Spectroscopy (USSR)
Teor. i Éksperim. Khim.—Theoretical and Experimental Chemistry
Theor. i Matem. Fiz.—Theoretical and Mathematical Physics
Zh. Éksperim. i Teor. Fiz.—Soviet Physics–JETP
Zh. Éksperim. i Teor. Fiz. Pis'ma—JETP Letters
Zh. Prikl. Khim.—Journal of Applied Chemistry of the USSR
Zh. Strukt. Khim.—Journal of Structural Chemistry
Zh. Vsesoyuzn. Khim. Ob-va—Mendeleev Chemistry Journal

Index